FIND THE LOST DOLLARS

DOLLARS

6 Steps to Increase Profits in Architecture, Engineering, and Environmental Firms

D1621855

June R. Jewell, CPA

Dedicated to my boyfriend Chris and my three kids, Jordan, Kyle, and Lauren, who have supported and loved me, and made it possible for me to accomplish my dream of writing a book.

PRAISE FOR FIND THE LOST DOLLARS

"June's book goes far beyond how to find the lost dollars. It includes primers on all the basics of running a good business, from marketing to client relations, and from cash management to computer systems. This is a must-read for any current or aspiring A&E firm leader."

– FRANK A. STASIOWSKI, FAIA,
Founder and President, PSMJ Resources, Inc.

"There is little to no business training in the architectural or engineering curricula, leaving most of us as we enter the profession woefully lacking in the business, sense, vocabulary, or skills necessary to guide our firms to financial success. Our clients, who depend on us to be stable, ongoing enterprises for the period of years during which we'll be working together expect us to be more than just surviving. June's book offers complete and detailed guidance that every architect and engineer needs to be successful."

– ED FRIEDRICHS,
Friedrichs Group, and former CEO, Gensler

"*Find the Lost Dollars* not only has relevant, current data and advice, but the last chapter gives you a gift: tools to assess your firm and how to find the lost dollars in your practice.....go find those dollars and be more profitable"

–TED MAZIEJKA, Consultant, ZweigWhite

"This book is truly a Jewel within the AEC business books that are available today. It is easy to read, full of real world ideas, samples, innovations, and ways to save (and make) more money now."

– RON WORTH, CAE, FSMPS, CPSM,
CEO, Society for Marketing Professional Services

"There is money hiding in the corners of your firm! June Jewell has pulled back the curtain and revealed the long list of lazy, outdated, and misguided business practices that conspire to keep AE firms from truly prospering. We've been suffering these bad habits for years, and now Jewell shines a bright light on them in a way that leaves no excuses. If you want higher profits, read this book!"

— **DAVID A. STONE**, CEO, Stone and Company

"Business authors often try too hard to show their smarts with gimmicks and clichés. You only need to read the first chapter of *Find the Lost Dollars* to see that June Jewell is an expert in the nature of operating a business for profit and quality. I didn't have to wade through theory and wasted words to get to the meaningful content. June is not just an author; she has done the job. She has been a consistent implementer of change. Having known June for a few years now I can tell you she is both tenacious and personable. She knows how to get work in the door and has a track record of providing true quality and measurable results. Her book provides a treasure map to finding your company's lost dollars with proof and easy to understand benchmarks. This book speaks our language."

— **CHRIS HILLMANN**, President/CEO, Hillmann Consulting, LLC

"*Find the Lost Dollars* is one of those business books that ties together all of the things that successful entrepreneurs know or need to know, and most importantly, provides a resource for succession. I plan to insure that 6-8 of our key staff receive the book, and encourage each one to read it and absorb the information—and of course, have them employ the knowledge on a daily basis. None of us in a CEO role today in A&E firms went to CEO school. This book will provide a short course and guidance to the next generation—those succeeding the current leadership—in our firm"

— **WES GUCKERT**, President, The Traffic Group

"If the ideas and action items provided in *Find the Lost Dollars* were implemented, not only the company implementing these ideas would be successful, but the entire mindset of the A&E industry would change!"

— **ZACK SHARIFF**, P.E., LEED® AP, CEO, Allen & Shariff

"*Find the Lost Dollars* provides exceptional perspective for the management and operations of professional services firms. It provided me with best practices and realistic suggestions that can actually be implemented."

— **TIM KLABUNDE,** Marketing Director, Timmons Group

"*Find the Lost Dollars* is a must read for anyone in the A&E industry, for the up-and-coming manager, financial staff, and principals. Even the most seasoned principals should be reviewing this material as a reminder of why these issues are important."

— **EDWARD (ED) J. KROMAN,** III, Director of Finance,
Summer Consultants, Inc.

"*Find the Lost Dollars* is a great reference for anyone from PM/Controller on up who wishes to be successful in this industry. All they need is the desire, will, and courage to change the way they operate."

— **ANTHONY (TONY) J. VITULLO,** CPA, CGMA, CFO,
Cooper, Robertson & Partners

"*Find the Lost Dollars* describes the fundamental business parameters affecting the building design industry. Executives who have the vision to train their future leaders before they put them into responsible positions, or as they place them into leadership roles, should make this book a must read for those ascending in their firm."

— **JOSEPH A. CAPPUCCIO,** Sr. Vice President,
Rolf Jensen & Associates, Inc.

"In the densely populated world of business management books, June Jewell's *Find the Lost Dollars* rises above the crowd, and distinguishes itself by virtue of the pragmatic, process-oriented guidance it provides, and identification of the eminently do-able corrective actions that we all need to implement in order to enhance (dare I say restore?) profitability. Well done June, and thank you!"

— **JAY APPLETON,** P.E., Principal, Kitchen & Associates

Contents

FOREWORD

In working with hundreds of architecture, engineering, and planning (AEP) and environmental consulting firms every year, I often see struggles with finding work, managing resources, and making money on projects. Growing an efficient and profitable businesses is a goal of most firm owners, yet the day-to-day issues of running the business often take precedence over doing what needs to be done to enable the firm to progress. Even the most successful firms, as represented at the annual Zweig Letter Hot Firm Conference, have at least one area of their business that does not perform at the level they want or expect.

Let me start by stating that I am NOT generally a fan of management books. There's just too much B.S. in most of them because they are written by people who are management book junkies, and they have no real frame of reference borne in practical experience. Not so with June Jewell in her book, *Find the Lost Dollars*. She really "gets" this business—

and sets the stage for understanding the cultural traps that have engulfed our industry—then shows readers specifically where money gets lost in their businesses and how to find it. By starting with a high-level look at the habits, processes, and activities that lead to lost opportunities, ineffective use of resources, and project overruns, she is able to accurately pinpoint the major areas where firms lose money. And that's half or more of the game we all have to play—not LOSING money.

This book is both an educational resource and a how-to guide for the AEP, or environmental firm owner and principal. Use the book as a tool. It will help you focus on the key areas you can improve and systematize to improve your bottom line.

June Jewell has been at this a long time, and her book reflects that intimate practical knowledge of our industry. In her more than 23 years of working with hundreds of A&E firms, both small and large, she has accumulated a wealth of knowledge and best practices that when applied to your business, can make a measurable difference in your results. In fact, when looking at the statistics that ZweigWhite has assembled each year in our annual Financial Performance Survey, they bear out the presumptions that June has made about the problems with many businesses, including the fact that many firms do not share adequate information with their managers and teams. In addition, the expert quotes, advice, and anecdotes included in the book from so many industry experts and firm leaders really add an additional level of depth to the advice that is provided.

This book offers a practical approach that you can use right away to implement new processes and systems in your business, and to measure the results. I recommend that you read the entire book first, and then use the guides, resources, and tools provided in Chapter 11 to implement the strategies recommended. I believe that with diligent and honest evaluation of your key metrics and business processes, you will see the issues in your business from a different point of view and be able to move your company forward.

~**Mark C. Zweig,** *Founder and CEO, ZweigWhite, Fayetteville, Ark.*

PREFACE

I was accidently "thrown" into the architecture and engineering (A&E) industry in 1989 when I worked for a CPA firm in charge of growing their management information systems (MIS) practice. My job was to help our clients determine their requirements and find the best accounting system for their business. Up to that point, I had worked with generic software systems that were focused on accounting features, functionality, and reporting.

But when one of our clients, Gorove/Slade, a successful transportation consulting firm in Washington, D.C., asked me to help them find a new system to manage their 50-person engineering practice, I soon learned that vertical software, designed specifically for the A&E industry, was a much better approach. The difference was in how projects are handled in the software and the need to understand profitability by project rather than just departments or profit centers, which is the focus of most generic accounting systems.

In the process of evaluating the top four or five systems on the market at that time, I helped them select the Wind2 system. We picked Wind2 because of its flexibility and deep focus on project accounting, billing, and employee profitability reporting. As a result of my experience with the implementation of Wind2 in 1989, I ended up leaving my CPA firm to become an independent reseller and consultant for the Wind2 system in the Washington, D.C. area. Over the following 15 years, I built a team of software and industry experts solely focused on working with A&E firms. Because of our unique location in the United States, we also became experts in government contracting, which crossed industries, including A&E, information technology (IT) consultants, defense contractors, and other project-based consulting firms.

In addition to back-office accounting, time and billing, and project management functions of the Wind2 system, we also worked with the AE Marketing Manager system that was later purchased by Wind2 Software. This was one of the original Client Relationship Management (CRM) products on the market specifically developed for A&E firms. It helped manage opportunities, and produced SF 254/255s. We also became experts in helping A&E firms manage their marketing and proposal efforts.

In 2005, Deltek bought Wind2 Software, and my firm, which was, at that time, named Jewell & Associates, became a Deltek partner. Deltek acquired the top three products on the market, Harper & Shuman's Advantage (formerly CFMS), Sema4, and Wind2 between 1996 and 2005. It had amassed market share and dominance in the industry with more than 12,000 architectural, engineering, and environmental firms as clients.

In 2000, Deltek developed a new and more technologically advanced system called Deltek Vision. What differentiates Deltek Vision from any other product on the market today is that it is a fully integrated, web-based enterprise resource planning (ERP) system built solely for project-based businesses.

What all of our clients had in common were projects. And in working with more than 700 firms over the last 22 years, I have come to develop recommendations for improving the business management practices of our clients and helping them to realize a return on investment (ROI) from their software systems.

My passion is to help our clients succeed by improving the way they manage employees, projects, processes, and technology. I have seen so many examples of poor business management over the years and the sometimes devastating effects that bad business decisions have caused for our clients.

One of the biggest mistakes I have seen is not leveraging technology to improve marketing, project management, and employee/resource management. Too often, the decision to buy software is based on the lowest price rather than how it can be a strategic tool to help the A&E firm owner to grow their business.

By taking the right approach to managing your business, you can make more money and have happier employees and clients. If you believe "how you have always done it" is better, I am hoping to show you that you can always improve and squeeze every last dollar from your business.

INTRODUCTION

For those of you who started your businesses more than 10 years ago, you have seen dramatic changes in what it takes to run a business and make a profit. For me, this has really hit home in the last five years, with the decline of the U.S. economy. I have seen clients with 100-year-old businesses file for bankruptcy. I have seen others sell for pennies on the dollar, as they failed to collect millions in past-due receivables. The A&E industries have changed, and we have had to change too. Many of the ways we used to run our businesses in the past are no longer working.

From how we go after business to hiring staff to managing our projects—almost everything is different from how it was 10 years ago. However, many firms have not made the shift and are still managing as they did 20 years ago. This is due to many reasons, including company cultures, outdated business management systems, as well as fear of the pain that it will cause to try and change.

Many A&E firms are started by design professionals who are trying to fulfill their passions in life and who deplore the business management aspects of running a firm. They were not necessarily educated to run a business and may have little to no experience doing so. Those that are successful are those who realize they are business owners first and architects, engineers, or designers second.

In his book, *Impact 2020,* Frank Stasiowski, FAIA, CEO, and Founder of PSMJ, comments on how the A&E industry is slow to move, despite the rapidly changing landscape of the world affecting the industry: "Compared to the pace of the change in the rest of the world, most design firms seem like they're slogging through mud. More than other professions, the design industry is too set in its ways."

Stasiowski suggests in his book that as a result of the rapid changes in technology, the climate, global expansion, and communication, hundreds of new opportunities are developing every day that will affect the industry. This book is a fascinating prediction of the changes in our industry that will determine what the typical A&E firm will look like in 2020.

Most A&E firms are not paying attention to many of these changes, and many have not even caught up with the technology the rest of the business world was using 10 years ago. Change is undoubtedly painful, but so is losing money! And even if your business is managing to get by, you put your firm at risk by not embracing the newer business models, technology, and best practices that today's most successful firms are employing.

For every process in your company, you have the potential to find lost dollars. These are the unbilled extra services, the proposals that never got sent, or the employee that is not optimally utilized. You can find these lost dollars by taking a different approach to your business and by examining your company's long-held cultural traps. The key is to be open to change and realize that "the way we've always done it" is not always the best way.

Our businesses are fueled by people, processes, and technology. Regardless of how successful you are, you can always improve. What works for a very small firm may not work for a larger firm with multiple offices and several levels of management hierarchy. Your processes must be scalable and need to adapt to changes in your business, such as the type of business pursued (discipline or public versus private), geography (remote locations), and contractual requirements. Today's virtual employees need different technology, communication, and processes to be successful. Competition is tough, and by focusing on a few areas that can have the biggest impact, you can realize sustained growth and improved profitability.

This book will take you through a journey of understanding the cultural traps your firm has embraced, and how they affect your success. We will look at where the lost dollars are hiding in your business, how to determine the magnitude of these losses, as well as how to develop processes and make the improvements necessary to recover the lost revenues or plug the leaking expenses.

These problems show up every day in your business as redundant data entry, lack of timely project information, inconsistent estimating and proposal practices, and hours spent on creating spreadsheets. Without these problems, reports are available easily and in real-time, accuracy is dramatically increased, and employees' time is spent more efficiently and effectively. The results you will get are improved management decisions, better project management, and ultimately, higher profitability.

Each area of your business processes can be improved. As we look at the life cycle of a project, we will evaluate each major process and determine where the biggest gains can be found. In Chapters 3 through 8, I will provide recommended practices and solutions for correcting the most prevalent problems in a typical A&E firm. While every firm is different, similar issues and patterns are certainly recognizable and fixable. I have worked with hundreds of firms over the years, both large and small, and lost dollars can be found in every firm. In Chapter 9, we will go into automated systems and best practices for selecting, implementing, and using them to minimize lost dollars and tighten up your firm's project management. In Chapter 10, we will look at the specific issues around bidding on and performing work for the government.

In Chapter 11, I show you how to actually locate the lost dollars in your business. You can jump ahead to read this first, but it will be much more meaningful if you have read the previous chapters. I will outline a 6-step plan for implementing the recommendations from the previous chapters and making the changes needed to move your firm toward greater success. In addition, you can go to http://www.FindTheLostDollars.com to download online tools that I developed to help you implement the 6 steps.

Through this process, and by making the changes you need to make in your business, you will also discover that your employees will be happier. No one enjoys frustrating, cumbersome, and redundant tasks that take up time at work and at the end of the day leaves them feeling like they never accomplished their goals. By making your staff members' lives easier, they will be more excited about their jobs and better able to contribute to the profitability of the firm.

Some people have suggested that this book is better suited as a training guide for A&E firm future leaders. I believe that the advice and best practices conveyed in this book should be the basis of education for all successful A&E firm executives and leaders.

If you are open to taking a realistic and penetrating look at all aspects of your firm's business management practices, you will find the lost dollars in your business, and be able to put the solutions in place to capture and increase your company's profits.

ACKNOWLEDGMENTS

I would like to thank all of the industry experts, thought leaders, and Acuity Business Solutions clients and staff that contributed their hard-earned expertise to this book. I could not have put together such a wide range of real-world advice without the quotes, stories, and other contributions of so many people who provided help in my mission to improve the business management of A&E firms.

Most of the principals and firm leaders who I appealed to for their contributions were very gracious, giving, and insightful. You will find some of their helpful gems scattered throughout each topic.

I am extremely grateful to Mark Zweig, a leading expert to the A&E industry and founder and CEO of ZweigWhite. He provided me access to his extensive resources including recent research providing detailed insight into the state of the industry today. Giving me the opportunity to contribute on a bi-monthly basis to the *Zweig Letter* has increased my exposure to many A&E firm leaders that are not my clients and allowed me to develop valuable content that is read by thousands of firm leaders every week.

Throughout this book, many other experts have contributed their wisdom, advice, and anecdotes. I offer special thanks to all of them for their insightful contributions (see Appendix B for a full list of contributors).

CULTURAL AND OPERATIONAL CHALLENGES IN THE A&E FIRM

CHAPTER 1

10 CULTURE TRAPS THAT AFFECT FIRM PROFITABILITY

"We cannot make a profit unless we take care of the customer, and we cannot take care of the customer unless we make a profit."

~GERARD ARPEY, FORMER CHAIRMAN, PRESIDENT, AND CEO OF AMR CORPORATION (PARENT COMPANY OF AMERICAN AIRLINES)

Many owners of A&E firms are not focused on making a profit. It is amazing how many principals smile and nod knowingly when I talk about this at industry events. Creative-minded design professionals often have a passion in life to design beautiful structures—not to manage projects and cash flow. Profit often takes a second, third, or last spot when it comes to getting the attention and concern of the A&E entrepreneur.

Profit needs to be a top priority in your firm to ensure its continued health, growth, and ability to attract employees. Profits fuel growth and allow you to achieve the owner's strategic objectives. Additionally, having a profitable business will enable you to provide better services to your clients and retain the best talent. Other important reasons to make a profit include giving you the ability to do the following:

- Improve your work space
- Compensate your employees better and attract more skilled employees
- Make it through tough economic or slow times
- Develop employee skills through training and other professional development
- Provide money for marketing so you can win more projects
- Invest in technology to increase productivity and create further profits
- Provide owners a return on investment (ROI) in the business
- Grow the business value for eventual transition
- Produce a win-win for both you and your clients!

A great anecdote on this topic, written by Ed Friedrichs, former CEO of Gensler, helps us understand how this long-held falsehood about profitability in the design industry has become so ingrained:

• •

My father spent a good deal of effort discouraging me from becoming an architect. "High-risk, low-reward," he and his pals said every time I talked about my fascination with designing buildings. Find a profession where you can make enough money to support a family and buy a house, where you can afford to put your kids through college. Being a dutiful son, I entered the university to study engineering. By the time I reached my senior year I had lost none of my interest in becoming an architect. I transferred into the architecture curriculum, much to my father's and my college counselor's chagrin. I credit my professional success with my stubborn focus on proving my father and his friends wrong. I became fascinated with designing business practices that made our firm highly profitable and instituted quality control systems that contained our liability.

• •

The culture of your firm directly impacts your profitability. How you view and spend money, deal with your financial management practices, and manage your staff directly translates into policies, processes, and behaviors that cause projects to go over budget. Without a healthy focus on profit, an owner may find it difficult to grow the business.

Many firms get trapped by their culture, keeping them from being able to grow and expand. These traps are long-held beliefs by the owners, often passed to them from previous employers or mentors. Being able to rise above these traps will enable your firm to reach new levels of success and growth, with false beliefs being replaced by techniques, strategies, and processes that represent the truth about the A&E industry. This will allow your company to realize the success seen by the top 10% of firms in the business.

The following are the main "culture traps" I see affecting how a firm manages money and influences its ability to be profitable. These mistruths cause many firms to ignore their contracts, budgets, and other project management best practices. By understanding how your firm embraces each of these belief systems and how it impacts your daily practices, you will be able to make the changes necessary to see real increases in company profitability.

This book will help you break down these obstacles, and help you view your firm with a new perspective. By focusing on the specific operational processes and deficiencies in your business management practices and measuring their impact, you will be able to locate the lost dollars caused by these traps, and incorporate new systems and processes that can have a measurable, positive impact on your bottom line.

TRAP 1: QUALITY IS EVERYTHING.

Most professionals want to offer the highest quality possible. However, the reality is most budgets are lower than the level of quality we desire. Trying to balance the level of quality with an inadequate budget is a critical, yet often poorly mastered component, of project management. If your firm consistently produces a work product that exceeds the level of quality specified in the contract, you will struggle to make money. Setting realistic expectations from the beginning is critical to a profitable project and a happy client.

If you are a firm that specializes in top-quality design features and materials, it is critical to find like-minded clients that are willing and able to afford it. The deadly combination of the belief that "quality is everything," combined with an unhealthy lack of focus on profit, will eventually kill the firm and destroy the possibility of success. This is an irony that will especially show itself in poor economic conditions, when design professionals have to pare back their standards to meet the decreasing level of funds available to complete projects.

The truth is that most design professionals consider their work "art." Their desire for recognition and awards is sometimes balanced against the firm's need to stay profitable. In the next chapter we will talk about scope creep and failure to bill for extra services. This trap is often one of the primary causes of scope creep. The failure to balance the design professional's desire for beautiful work with a less-than-adequate budget and client synergy is a deep-rooted issue in the A&E industry.

TRAP 2: KEEP THE CLIENT HAPPY AT ALL COSTS.

We have all heard the old adage, "The customer is always right!" However, this trap can have the unwanted result of derailing a company's project management practices to its financial detriment. The truth is keeping clients happy throughout the duration of a project can be challenging and costly. In our desire to have a happy client, our employees may engage in practices that threaten the profitability of the project. It is critical to the success of a company that employees learn how to set client expectations, communicate often and well, and deal with conflict as it arises during the project.

A good client is one who is fair and does not take advantage of our desire to keep them happy at all times. Your staff will need guidelines for dealing with overly demanding clients who change their minds or ask for out-of-scope changes. Having a senior manager involved in these scenarios can help keep the project on track and serve to train junior staff in how to handle these difficult situations.

Christine Brack, Principal at ZweigWhite, Fayetteville, Ark., has an excellent example of how this trap can lead to decreases in profits and even losses on projects:

I worked with an engineering firm to analyze the amount of work they gave away as small favors or allowed to slide as scope creep. **In one calendar year we accounted for $250,000 they spent but never billed.** *The firm's philosophy was certainly to take care of the client, but across seven project managers, this turned out to be a hefty sum. These built up over time and were innocently small numbers with each occurrence, but the seriousness in the end is that this was a quarter of a million dollars not invested back into training and not given as bonuses. As a result, they were more diligent about writing scopes and proposals, and kept a careful watch on any creep from that point forward.*

Does your firm engage in this type of practice? This problem is definitely more prevalent today as work is harder to get, and keeping clients happy is one of our highest priorities. One of the main areas where this trap shows up is having many more meetings than were planned for in the budget. When a client calls, we jump, rarely questioning whether the work being requested is in the scope. In some cases, our employees actually ignore the scope, believing they are doing the right thing by giving the client everything they ask for. Managing this delicate balance of responsiveness versus fiscal responsibility to control costs on the project takes skill, experience, and attention to detail.

Trap 3: In slow times, it is OK to take projects we know will lose money.

With the downturn in the economy and especially the A&E industry over the last five years, many firms have struggled to stay in business and retain staff. As competitors continue to bid lower and lower, and larger firms start competing more on smaller projects, most companies feel pressure to lower fees to a level that they cannot permanently sustain. As a result, many firms feel compelled to bid on projects at a lower cost than they can do the work, knowing they will lose money. The reasons for this are usually centered around keeping staff busy so that layoffs won't be required or getting in the door with a new client.

While this is understandable and occasionally necessary, there are several issues with this approach. First and foremost, you are setting your company up for a long future of losing money with that client. You can rarely start off at a low threshold and increase it later. In most cases, you will never be able to get a profitable project with this client, and you will get trapped in a long-term vicious cycle of continuing to provide services at a loss. Very often, your client will continue to take advantage of this long after the economy has improved.

Another reason this is a bad idea is that it can have the negative effect of causing tension between your project managers (PMs) and the client, as they struggle to keep costs in control and limit the amount of loss that is realized. This practice leads to a negative relationship with the client, and an unfair burden placed on your PMs, as they strive to make the client happy (Trap 2) *and* help the firm make a profit.

A third reason is that this practice has a negative impact on the industry as a whole, which is difficult to repair when the economy improves. There has been a trend over the last five years toward declining fees, and this practice only leads to pushing them lower for all of us.

Bob Gillcrist, A&E Industry Consultant, provides this insight into this challenging dilemma:

> *Loss leaders do have their place, but I always caution people not to take a "loss leader" just to get work in the door. It is something that should be more strategic, like getting into a new market or client that opens doors to future work, and looking at it in the financial context of the overall health of the company.*

About four years ago, I worked with the owners of a small regional engineering firm that was really feeling the pinch of the economic downturn. They had already let go of a few staff members and were really trying to hold on to the rest of their staff. They made the decision to lowball a bid on a large project just to get some work and pick up a new client. They won the project, with a developer who was barely making it as well. This project was one of the most painful projects on which they had ever worked. They constantly struggled to keep the project within scope, as well as on time and budget. The client ended up cutting corners whenever possible, and they often had a hard time getting paid. Today, the company is still making up for the losses on that project and is committed to not going after loss leaders again.

In analyzing your project profitability margins, I recommend that you take a good look at whether you are bidding too low on projects as a rule or are trapped working for clients who require too much for too little. Determining the areas of true profit within the company can help you develop a strategic plan that focuses on the key areas in which the firm is more successful.

TRAP 4: ALL CLIENTS ARE GOOD CLIENTS.

We all know intuitively that all clients are not good clients, yet we continue to work with clients who are cheap, treat our staff poorly, and do not appreciate the work we do for them. Often times, you may love the project, and desire the profits and exposure that the project affords, yet dread working with the client for a host of reasons.

Other times, a bad client is one who changes his or her mind all the time, is too demanding, or does not pay on time. We have become conditioned in recent years to accept bad behavior from our clients. We allow them to drag us into too many meetings, change their minds all the time, and then pay in 90-day terms.

The truth is that life is too short to work in these conditions! I know this is a bold statement, but ultimately if you are not happy, then I can bet that your employees are not either. And studies show that if employees are not happy, they will not perform at their best, which continues to erode profits and client satisfaction.

I recommend you look at categorizing your clients into "A," "B," and "C" clients. Determine the specific criteria that are indicative of your ideal or "A" client, and figure out a way to "measure" each client according to these standards. Your "A" clients are who you want to ensure you have a solid relationship and focus working with more. By figuring out who is your ideal client, and distinguishing the best clients to work with and who provide the highest return for the firm, you can guide your company to an overall more profitable position.

TRAP 5: YOU CAN'T LOSE MONEY ON A TIME AND MATERIALS (T&M) CONTRACT.

If you do a lot of T&M contracts, it is really important to understand this trap. Many A&E firm leaders and PMs believe that you cannot lose money on a T&M contract. This is assumed because you are able to bill for every hour that you work, and unless there are write-offs of hours that are not billed, you are making money.

The problem lies in how the labor billing rate is formulated. A bill rate is calculated by adding all the direct costs of the labor, plus any direct personnel expenses (DPE) such as fringe benefits, as well as enough to cover overhead, general, and administrative (G&A) expenses and a desired profit margin.

The issue with T&M contracts is that in order to make the target profit, or the actual fringe, G&A and overhead rates must be in line with the estimated rates. If the actual rates are higher than the estimated rates by the amount of the target profit percentage, the company loses money.

This can happen with little or no fault of the PM and may not be calculated often enough to make necessary adjustments. Since many T&M contracts have fixed rates, with no ability to change them annually, your firm could be stuck with an unprofitable T&M contract for an extended period of time. Only by controlling the overhead rate, and ensuring that your billing rates are calculated correctly, are you able to make a profit on these projects.

Another more widely known risk with T&M contracts is the issue of rework. If poor-quality work is performed, or the work is not in line with the client's expectations, it may have to be redone. This rework should be tracked separately so that staffing or quality problems can be addressed.

One further issue with T&M contracts alluded to earlier in this text is time that is held or written off and never billed. Firms that allow this to consistently happen will lose money on T&M contracts, and utilization reports may be distorted. Measures should be put in place to monitor and correct these practices, especially if you are not making your expected profit margins on T&M contracts.

TRAP 6: WE DO NOT SHARE FINANCIAL DATA WITH MANAGERS AND EMPLOYEES.

Sharing financial data with your PMs is a key management decision that affects how successful they can be in controlling project profitability. Many project management experts argue that the more they know, the easier it is for them to react in a timely manner to project issues and overruns.

Many firm executives are concerned about managers and employees knowing too much about how much others make or how much profit the firm makes. There must be a balance between keeping certain information confidential, such as executive compensation plans, and giving them the data they need to manage their projects.

One concern about not sharing this data is that if managers and employees don't know what is really going on they will make their own assumptions about how the company is doing. They may assume that the owners are making much more than they are and feel resentful if bonuses are not given, or are lower than they think they should be. By sharing the true cost of running the business, employees have a better feel for the struggles of the owners to pay the bills and collect the receivables, as well as the risk they are taking. This can lead to a better motivated workforce that is more understanding of the decisions that management makes.

In a January 2000 *Business Week* article entitled, *Keep Employees in the Dark, and They'll Go Where It's Light,* the paradox of sharing information with employees is discussed. This article argues that while management may feel they are keeping employees from making false assumptions and panicking, the lack of information causes employees to imagine the worst, and may drive them to go look for another job. (Source: http://www.businessweek.com/smallbiz/0001/bk000114.htm)

Derrick Smith, Senior Vice President of Mackay Sposito in Vancouver, Wash., highlights his company's success with openly sharing his firm's financial data with its employees:

• •

During the economic downturn, the partners at MacKay Sposito made a decision that we would openly share our company's financial situation with our employees. This represented a significant change in the way we managed our business. It also changed our culture. We found that employees became more invested in our company's success. Everyone felt like we were "in-the-boat" together. It was a leap of faith that drove our financial success and has now become a cornerstone to the way we run our business.

＊ ＊

There is a lot written on this topic, and each firm must find the right level of transparency with which they are comfortable. The recent thinking in this area is that more is better. Managers will feel more empowered and motivated to help the company succeed if they feel they are trusted and have a direct impact on the success of the firm.

TRAP 7: OUR CLIENT DOES NOT WANT US TO MAKE A PROFIT.

The design and construction world is under pressure these days to cut costs and streamline budgets, and our clients are pushing harder for lower fees. This pressure has created a common myth in the industry that our clients do not want us to make money, and we should bid as low as possible.

It is critical to our short- and long-term success that we strive to find the clients that truly value our work and understand how important it is to the success of their projects that we make a profit. Operating at a loss causes us to have to cut back on quality, using cheaper labor than we would normally recommend and also possibly making critical decisions based only on cost. While this may be the reality in today's world, it is ultimately a threat to the success of our projects and our firms.

A good client does have a win-win attitude and knows that it is in his or her best interest for us to make a profit. For all the reasons I described above, and especially our ability to attract and retain talented staff, our client's project can only be as successful as we are. A losing project can strain our relationship with our clients and cause our employees to resent their work. Focusing on the "A" clients, as described previously, will ensure that we are working with people who value our services and treat us with respect and fairness.

John Saber, President and CEO of Encon Group in Kensington, Md., illustrates this point beautifully with this inspiring story:

• •

Several years ago, a client of ours sunk his entire net worth and then some into converting an old building into mixed retail space and condominiums. The project went horribly wrong, and we lost an incredible amount of money. If fault had to be assigned, it was everyone's fault. We worked our way through it to completion and never tried to collect a small amount of money that was owed to us. Two years after completion, I received a check for the outstanding amount owed. I was shocked, so I called him and asked why he paid me. He told me that "out of everyone involved in the project, you were the only one who stood by me to the end." I've never had a prouder moment as business owner, both for me and my staff.

• •

TRAP 8: WE CAN'T MAKE OUR EMPLOYEES FOLLOW OUR POLICIES.

As firms "grow up," they usually find it necessary to begin implementing more internal controls, policies and approval processes in order to effectively manage people, as well as to execute on their project delivery and financial management. However, despite the fact that they put these policies in place, many firms do not enforce compliance with their policies, and several key aspects of firm financial management are compromised.

In working with hundreds of A&E firms, I have seen this trap more frequently than all the others combined. This is truly a cultural issue that goes back to the management's desire to take it easy on their staff and not micromanage people, and also to avoid a culture of bureaucracy.

One of the most detrimental results of this trap is poor timesheet practices, which can have a profoundly adverse effect on revenue, cash flow, and ability to charge for extra services. This practice can also create a bottleneck in the company that can limit growth and prohibit change.

Companies need good management policies to maximize project success, and failing to control critical business functions through necessary policies and processes will erode a firm's ability to achieve its goals. In Chapter 6, we will specifically look at timesheet practices and how the tightening and enforcement of your time management policies can make a huge impact on the profitability of your firm.

TRAP 9: TIMES ARE TOUGH, SO WE CAN'T SPEND MONEY.

If you have been in business for more than 10 years, you have probably lived through a recession or two. It is inevitable that the economy will have ups and downs and our businesses will face financial challenges. Most firms took the necessary steps between 2007 and 2009 to trim the fat from their overhead, do the painful but necessary layoffs, and hunker down to weather the storm.

But the truth is that taking this defensive position will not work forever. As I publish this book in 2013, the future of the economy is unknown. In some ways this might be good, as we can proceed with cautious optimism rather than the crazy growth and spending days we experienced in the early to middle 2000s. That was a time of living big without a realistic view of the future. Now we can focus on growing our businesses so that we have a future to look forward to.

The lesson here is that we need to invest the money we make in the good times to get through the tough times. You have a choice to make: Should you invest in your business in order to pursue growth or stay as you are? With interest rates this low, it is difficult to find good investments. Our businesses, with potential returns as high as we can drive it, are still the best investment we can make, and we are responsible for our own destiny, rather than the finicky and risky stock market.

This is the time to take a hard look at your business and see where you can drive real efficiency, and generate bigger margins on your projects. Everything is driven by people, processes, and technology, and it is up to you to invest in the right areas to help your company achieve its goals and remain competitive.

Recently, at the Zweig Letter 2012 Hot Firm conference in Aspen, Colorado, a panel of CEOs was asked about their strategy for succeeding during the recession of the last five years. In almost every case, the CEO stated that they had increased their marketing budget and focused on increasing their win rate. I believe this is a great strategy for both enduring a downturn, and setting up the firm to succeed when the economy begins to recover. The firms that are doing well now are the ones that either continued to invest or increased their investment in their firm's future during the recession.

TRAP 10: THIS IS HOW WE'VE ALWAYS DONE IT.

If this is your answer for the way you do anything in your company, then it's very likely that your processes are outdated and inefficient. Your employees are given guidance as to how best to manage their time, deal with clients, process transactions, and communicate. If they are doing these things the same way that they were doing them a few years ago, then your company is not maximizing the technology available today to save time and improve efficiency.

This trap often gets perpetuated because generations of employees have trained new staff, and continued to pass down the same processes as they were taught. If you ask them why they are doing something a certain way, they will often tell you that they don't know—that is the way they were trained.

Anthony (Tony) Vitullo, CFO at Cooper Robertson in New York City, reinforces my observations with his research on this phenomenon:

• •

In my years of experience, I have seen the perception of self-success prevent many business owners from taking good advice and making the right changes to successfully operate, grow, and sustain their business, even though that acquired success could have been greater with a strategic change here or there. Change comes hard, and those that embrace strategic change at every turn build a business to last, and are continually successful, those that don't languish in mediocrity or die. I experienced this in face to face interviews with entrepreneurs in the research I performed for my doctoral dissertation, and have seen it throughout my career. Successful owners know their business and their markets, and change with it even if it is ever so slightly (often as their markets change at that same pace).

• •

If your firm is not constantly looking at improving and taking advantage of new technology that is available for business and project management, then you are going to lose your competitive edge. No matter how much better your people are than your competitors, if they are not given the tools and training to ensure that their work is as effective and efficient as possible, they are wasting their time and your money.

This book will dig deep into how revamping your processes can make your employees more effective, and save valuable time and money on your projects. We will also explore how improving your computer systems can have a major impact on project success, client satisfaction, and employee satisfaction.

Summary

There are 10 cultural traps or myths that I commonly see in A&E firms that, when applied to daily operations, tend to erode profits. These 10 traps are as follows:

Trap 1: Quality is everything. Firms focus on the quality of the final product without regard to the client's budget or the scope of the contract.

Trap 2: Keep the client happy at all costs. Companies do whatever the clients asks, also without regard to the budget or scope.

Trap 3: In slow times, it is OK to take projects we know will lose money. This is a losing battle that can cause you to get "stuck" in a long-term relationship with a client who expects you to lose money.

Trap 4: All clients are good clients. Firms fail to see how they should not work with some clients. I recommend you categorize clients into "A," "B," and "C" based on several criteria.

Trap 5: You can't lose money on a T&M contract. Overhead can impact the profitability of a T&M contract. Also, write-offs and rework can have a big impact on profits if not controlled.

Trap 6: We do not share financial data with managers and employees. Theories differ on this topic, but most experts agree that many firms do not give managers and employees enough information about how the firm is doing. This can have a negative impact on performance.

Trap 7: Our client does not want us to make a profit. Despite common belief, most clients do want you to stay healthy and not struggle to get by, especially while working on their projects.

Trap 8: We can't make our employees follow our policies. Policies are necessary to ensure that the firm can grow and deliver a predictable service, yet many firms struggle to enforce compliance with company policies. This can have a very detrimental effect on profitability.

Trap 9: Times are tough so we can't spend any money. Cutting back on expenses during tough times can have a detrimental effect on the firm's ability to grow and hire the best employees.

Trap 10: This is how we've always done it. If your firm is not keeping up with technology and is refusing to improve, you run the risk of being beaten by your competitors. Change is inevitable, and if implemented wisely, it can be a great boost to financial success.

CHAPTER - 2

NINE AREAS WHERE MONEY GETS LOST IN THE A&E FIRM

"The successful man is the one who finds out what is the matter with his business before his competitors do."

~ROY L. SMITH, AMERICAN CLERGYMAN AND AUTHOR

As our businesses begin to grow, and we become more successful at winning work and hiring employees, we start to develop processes to handle all of the direct (project-related) and indirect (overhead and G&A) functions of the business. These processes involve policies and procedures, systems, communication methods, and levels of structure and accountability.

If we are growing, we usually realize that we must change, become more structured, and build out an infrastructure to support our changing business. This may happen many times in a company's life cycle, and require annual review of how projects, people, and financial practices are managed.

Some design firm owners and their employees will fight this change, and a business can begin to experience some conflict as the necessary processes to execute and make money on projects are implemented. While some firms handle this evolution with minimal disruption and drama, other firms will find that they are constantly struggling to get employees to follow their processes, and project performance can suffer.

Where Is Money Hiding in Your Business?

By taking a look at the nine areas where money may be hiding in your business, you can find ways to address the culture traps affecting your project profitability and discover ways to turn around your projects to positive performance.

We will explore some specific steps for addressing each of these areas later in this book, but as you read through the next few pages, use the formulas provided to assess which of these areas are your biggest problems. You will not be able to address all these challenges at once. In Chapter 11, I will help you determine which ones will make the biggest positive impact on your business and develop a strategy to attack it.

In order to help you calculate the lost dollars in your business, I have provided some metrics for comparing your firm to industry averages (available as of Q4 2012) and assessing the potential impact you can make in your business by addressing each issue. These metrics are found in the *ZweigWhite Annual Financial Performance Survey for 2012*. The important thing here is to think about change. How would a 1% change in each metric impact your financial success? I will help you calculate that so you will know which areas to focus on first.

For each of the areas addressed below, you may be losing revenue, or incurring unnecessary costs. To be able to turn around each area, you need to assess where the leaks are and make changes to your processes and systems to address them. After you have evaluated which areas are your primary pain points, you can use the suggestions in the chapters that follow to make the changes needed with your people, processes, and systems to find the lost dollars in your business.

It may cost money to solve these problems. You need to ask yourself, would I spend $20,000 to save $50,000? In many cases, a simple investment in systems can make a big difference. Many small business owners view company expenses for key assets and infrastructure as necessary expenses rather than strategic investments. Decisions to spend money in a business should be based on an assessment of several factors, including ROI, strategic positioning, and elimination of problems that are holding the company back from achieving its goals.

Ultimately the product, service, or employee being invested in must produce something valuable for the company, and determining this value is the key to deciding whether or not an investment is a good one. By determining the value of the lost revenue or the wasted costs, you can start to assess the value of fixing the problem. Use the calculation estimates provided in this chapter to gain an understanding of the size of the problem in your firm. In the chapters that follow, we will look at some solutions to these problems.

Cost of Lost Opportunities

In looking at the life cycle of an A&E project, it all starts with an opportunity. There are some opportunities that we win and some that we lose. As we go through the process of investing money into pursuits, there is often a "Go/No-Go" process that includes a series of questions we must analyze when deciding whether to bid on and execute a successful project.

So what does it cost us to lose a potential project? For the most part, it is the lost revenues and profit from the project itself. However, we are also losing our investment in the proposal process, the cost of not having our staff at full utilization, and the portion of the project revenue that is covering our estimated overhead.

Improving on this process will not only save money and time used to bid on a losing project, but also inherently affect your cash flow. In looking at the project life cycle, the amount of time that elapses from when you bid on a project until you bill and collect for your work can be six to 12 months or longer! This can force a firm to have to borrow money to cover the full 50% of the cycle that does not produce cash. This added cost of borrowing should be another factor in the analysis of whether to go after a project. You need to ask the question: **Can we afford to finance the proposal effort based on the probability that we will win?**

If your firm is bidding on many different projects at the same time, doing this analysis and tracking the success of all of your proposal efforts can be daunting. With managing the resources required to ensure that estimates are prepared to ensure profitability, developing proposals with the required winning components, and tracking all of the essential milestones and deadlines, it is no wonder lost dollars are hidden, trapped, or lost throughout the process. A good client relationship management (CRM) system can help manage and control many of these processes. In Chapter 4, we will look at ways that a CRM system can help you find the lost dollars hiding in your business development processes.

When considering the cost of missing just one critical deadline on a major proposal effort, it is easy to see that effective management of this process will not only save the firm money, but also ensure a higher level of success in winning work.

In addition to the risk of losing a project in the bidding process, there is also a cost to winning a project on which the company will lose money, or is not suited for the company. Bob Gillcrist, industry consultant and former executive at HOK, offers this insight as to the importance of the Go/No-Go decision and its impact on the firm's success:

• •

A well-considered Go/No-Go can provide a critical first step in establishing points to be evaluated by not only marketing people but operations/management people to determine if a project has the potential to be either successful or a loser. The initial Go/No-Go may indicate that this should be a project to pursue, however, during the fee proposal stage or the contracts term phase there may be issues that cause you to walk away from the project. So, Go/No-Go is not only a crucial process to appropriate project pursuit, but does not stop at the initial assessment of the opportunity.

• •

In Chapter 4, we will look at some best practices for evaluating the Go/No-Go process and managing opportunities for better pipeline analysis and deadline tracking.

The **win rate** is the primary metric used to determine the success of your marketing efforts. In looking at the number of proposals won versus the number submitted, you can start to understand the amount of money that could potentially be found from improving your opportunity management processes.

Throughout this book, I will use a simplified example of a $10 million design firm. The following calculation is an example of the cost savings from managing opportunities better and utilizing an effective Go/No-Go process.

Savings from Improving the Opportunity Management Process	
Total revenue	$10,000,000
Number projects proposals submitted	100
Number of proposals won	30
Win rate	30%
Average proposal size	$370,000
Amount realized by attaining a 1% higher win rate	$370,000
2012 AEC average industry win rate*	37.5%

ZweigWhite 2012 Marketing Survey

Lost Revenue from a Cumbersome Proposal Process

In addition to the lost revenue from a poor opportunity management process, the win rate is also impacted by the effectiveness of our proposals. To build on the previous discussion on the cost of investing in lost proposals, many firms are just not able to bid on as many proposals as they would like because of a cumbersome proposal process. This is often due to inefficient processes, access to project and employee qualifications data, and reliance on key PMs to do the heavy lifting of writing the proposals. In addition, this lack of access to critical data can affect the quality of the proposal in how well it is targeted for the type of project you are pursuing.

We often see that proposal coordinators and other staff are reinventing the wheel every time they have to create a new proposal. They spend a great deal of time looking for project and employee data, as well as messing around with proposal formatting and production. This inefficiency also causes communication strain between the marketing and accounting departments, when optimally, you would want cohesive teamwork. Accounting staff is often interrupted to provide financial data to the marketing staff, and marketing staff is frustrated because the information is not available as fast as needed to make deadlines.

All of this inefficiency is exacerbated if you are submitting government proposals including SF 330s or 254/255 forms. These forms can be quite time-consuming to prepare and require that a great deal of data are checked against other data for accuracy. Unfortunately, many of my clients have told me that they do not have the database or time available to be able to accurately determine the data for government forms, so they end up "guesstimating" the correct numbers.

The end result of this cumbersome process is that the quality and quantity of proposals your marketing team produces is not optimized. So what is the lost revenue or cost from an inefficient proposal process? This can be calculated by looking at your average proposal value and determining the revenue lost for each proposal you are not able to submit. Another way to look at this is what you could potentially bid on and win if you did not have the inefficiencies that are limiting your output.

Savings from Improved Proposal Process	
Number of proposals produced/year	100
Average amount of time to produce one proposal	120 man-hours
Average burdened rate per hour of proposal staff	$35/hour
Average proposal value	$370,000
Cost saving for 10% savings in proposal time	$42,000*

*This is just the savings from improved efficiency. Improving your cumbersome process can also have a significant impact on the number of proposals able to be prepared and the quality of the proposals, which can impact the win rate calculated above.

Lost Revenue Due to a Flawed Estimating Process

Creating good estimates is the key to success in realizing a profitable project. The estimate ultimately dictates the scope of the contract, budget, and process for delivering the project. If there are any flaws in your estimating process, you will experience losses on the project from budget overruns. The major problems that I commonly see in the estimating process are:

- Lack of consistency between estimators
- Inadequate level of detail in how hours are estimated by phase or task
- Using old or inaccurate information
- Failure to build in a margin for contingencies
- Using unstructured processes or lack of systems for developing and calculating estimates

Since all projects start with an estimate, effective processes must be put in place to ensure that the rates, structure, and assumptions used to calculate the estimate are valid and accurate. Managers tend to be extremely busy and pulled in many directions, so short cuts are often taken to streamline the process. By putting better processes and systems in place to prevent these oversights, you will start each project with a better financial "blueprint" for ensuring its success. Chapter 4 details some best practices for estimating and budgeting a project to produce the desired profit margin.

In figuring the cost of a poor estimating process, you must investigate the potential budget overruns that can occur, as well as the cost of potential scope creep and unrecovered extra services. There is also inherent risk in the estimating process, as a bad estimate will potentially lead to a bad scope and contract. Many A&E firm owners are shocked when they realize the effect of these numbers on their profitability. Because estimates and budgets directly affect the project profit margin, we will ultimately examine them together, in terms of the impact of improving the processes around them.

Calculating the Cost of Bad Estimating Processes	
Number of proposals per year	100
Number of projects won per year	30
Average estimate amount	$370,000
Percentage of projects that go over budget	20%
Percentage amount that projects go over budget	15%
Amount of average budget overrun	$55,500
Total value of estimates won	$11,100,000
Value of 1% improvement in estimating	$111,000

Lost Revenue and Excess Costs from Scope Creep or Extra Services That Are Not Recovered

As we just discussed, the key to effective project management starts with the estimate. The project scope is a critical factor in this estimate, and not only determines the process we will use to execute the contract, but also dictates the phases and tasks that make up the plan, and the resources needed to carry it out.

In order to ensure maximum profitability on a project, your team must execute on the project at or below the thresholds inherent in the plan. Any variation from the scope, including using different people or labor categories than were planned for, taking more time to do the work, or performing tasks that are outside the scope, will result in reduced or negative profit margins on the project.

Effective communication, planning, and monitoring have to be inherent to your process in order to ensure that your projects are executed according to the plan. This requires that you develop internal processes to set appropriate client expectations, communicate the scope to the project team, scrutinize timesheet accuracy, and monitor staff assignments and adherence to the scope.

Because of the traps we discussed in Chapter 1, such as "quality is everything" and "keep the client happy at all costs," employees and managers may find it easy to justify a departure from the project scope. This can become a consistent habit that over time will erode profit margins and ultimately the firm's ability to finance itself.

According to the *2012 ZweigWhite Project Management Survey*, 55% of A&E firm principals say they always share the complete scope of services with their PMs, and 40% say they only sometimes share the scope. For me, these are scary numbers and much worse than I imagined they would be.

I consistently identify the following problems in A&E firms' practices regarding managing the scope and control of extra services:

- Lack of policies and process around managing and controlling extra services requests
- Failure to adequately communicate the scope to the project team
- Poor timesheet practices
- Inadequate discussion and documentation around client expectations
- Failure to document how out-of-scope requests are to be handled in the contract

A focus on scope creep and controlling of extra services can provide a significant increase in project revenues. Chapter 5 takes a look at some practical tips for recognizing and billing for all of your extra services. So what does scope creep cost your firm?

Calculating the Lost Dollars from Scope Creep	
Number of projects worked on per year	40
Amount of average project billed per year	$250,000
Percentage of extra services not recovered	7% per project (estimated)
Lost dollars from extra services not recovered	$17,500 per project
Total amount lost on all projects	$700,000
Savings from 1% increase in extra services billing per year (40 X $250,000 X .01)	$100,000

Cost of Low Utilization and Poor Resource Scheduling

For a business that effectively sells people's time for a living, your employees and their time are your biggest assets. Managing these assets is crucial to the financial success in your A&E firm, yet the nature of the business is such that this is often the most difficult factor to control. Utilization is the main metric that indicates how effectively your firm is managing your human resources. Target utilization rates need to be set for each employee and monitored against actual performance.

Two types of analysis must be in place in looking at employee utilization. The first and easiest is reviewing the past. We can look at hours charged to determine where we are financially on our projects and make adjustments based on the work estimated to complete (ETC) it.

It is also critical to be able to forecast expected hours to be charged in the future. This is based on work remaining, availability of key team members, and expected requirements of other projects that are currently in progress or expected to begin. This requires a careful balancing act similar to spinning multiple plates in the air with an end goal of keeping all of your staff working at target utilization rates or higher.

In contrast, most firms are striving for a utilization rate higher than what they are actually achieving. For a firm to consistently hit or exceed its target, it must put in place processes and systems to help it stay on top of changing schedules, project requirements, delays, and employee availability issues. Being able to forecast can ensure that informed hiring and layoff decisions occur.

Not having this ability to predict the future can cost the A&E firm a great deal in terms of lost revenue, budget overruns, missed deadlines, and underutilized staff. Low utilization is the number-one cause of overhead escalation, which has a resulting negative impact on project profitability, and eventually, cash flow. By recognizing shortfalls in your employee time management, scheduling, and forecasting processes, you can begin to determine the cost of ineffective labor management.

Your firm cannot endure for long if this serious issue is not addressed. Matching the right amount of work with the right number of billable employees is both an art and a science. Your ability to address any deficiencies your firm has in this area can provide a huge payback. Improving your resource management, along with a tight time management process, should produce a major difference in your project profit margins.

Robert Skepton, Chief Financial Officer at Hillis-Carnes Engineering Associates in Annapolis Junction, Md., talks about the importance of project and revenue forecasting in today's post-recession economy:

• •

The focus on, and the expectation of precision in forecasting has changed dramatically since the recession of 2008. Pre-2008, there was an overriding sense that while there has always been significant seasonality in our business, the general trend, minus some very short term recessions, has always been and always will be up. Now with the significant uncertainty in the U.S. and global economies, and so many in our industry seemingly "waiting for other shoe to drop" with the U.S. debt issues, gaining an additional month or two of foresight into revenue is perceived to be an important strategic business tool. While you never have quite enough information as you would like to have when making forecasts, the importance of the continued tweaking and improving of your forecasting methods given what you do have should not be underestimated.

• •

Calculating the Cost of Poor Resource Management	
Number of full-time billable resources	70
Average billing rate per billable resource	$110
Total billable hours	87,360
Total standard hours	145,600 (70 x 2,080)
Company average utilization rate	60% (87,360 ÷145,600)
Value of a 1% increase in utilization per year ([145,600 x .01] x $110)	$160,160
Average industry utilization rate ranges between 55% and 74%	

*Depending on the type of work, size of firm, and region of the country per *ZweigWhite 2012 Financial Performance Survey.* Utilization rate, also called chargeability ratio, can be calculated using dollars or hours.

Cost of Poor Project Management

We rely on our PMs to drive the core functions of our business and its resulting profitability. If managers are not able to bring projects in at or below budget, the firm will lose money. The reasons why managers struggle with financial success on projects are varied. I have detailed the following three major categories of issues that you can address to understand why your projects go over budget. Chapter 5 will address some things that can be done to correct these three issues.:

- Poor project budgets
- Failure to use resources as planned
- Unexpected changes in the project scope, resources, or outside factors

Poor Project Budgets

As we discussed earlier, the estimate is the first step to producing a viable and executable project plan and budget. If the estimate is flawed, the resulting project budget will most likely be flawed as well. The estimate is usually developed by determining the phases and tasks to be accomplished, a high-level assessment of the labor categories to be assigned to each phase, and the estimated hours and dollars needed to complete each phase or task. In many firms, this is done using spreadsheets and rough, rounded rates for each labor category.

Once a project is won, the estimate should be converted to a more detailed project plan and budget. At this point, the labor categories may be converted to actual people, with more specific rates and hours assigned along a timeline. A good budget contains all of the following crucial components:

- A work breakdown structure that includes all of the phases and tasks in the order to be completed
- A start and end date for each phase and task
- The people or labor categories to be assigned to each phase and task at budgeted rates
- The estimated hours spread along the timeline. The closer the hours can be assigned to the timeframe in which they are expected to be spent, the better.
- The expenses expected to be incurred; also by pay period or month
- Subcontractor costs expected to be incurred by phase, task, and month

29

- Line items for additional services added to the scope after the project is started
- A line item for contingencies
- An estimate for meetings within each phase

These last two items are critical to preparing a realistic budget, yet many PMs fail to add line items for contingencies and meetings to their estimates and budgets. Contingencies can include staff turnover, client changes, project delays, unexpected regulatory issues, and other factors that can impede the progress and increase the cost of the project. Certainly, meetings are not unexpected; however, most managers either fail to include line items for meetings or severely underestimate the number of meetings that will be required. Considering this, it is no wonder that too many projects go over budget.

Failure to Use Resources as Planned

After the baseline (initial) budget is created, the hard part starts. This is where your PMs have the tough job of juggling the available resources with the project schedule and budget. Because of all the traps and challenges we discussed above, in addition to the fact that PMs are rarely trained in how to do all this, it is a constant battle to achieve both a profitable project and high utilization.

These two variables, project profitability and utilization, are often directly opposed to each other. If a PM keeps utilization high, then projects are very often pushed over budget. Knowing they are being measured by their utilization, employees are likely to charge their time to fixed fee projects in order to look billable. This serves to push up the project costs and lower the profit margin on the projects. Finding that balance often feels like performing a magic act!

In addition to the challenge of keeping people busy and allocated to the right projects at the right time, the PM must stay focused on getting the work done correctly, on time and to the client's satisfaction. Most firms focus on this, rather than the profitability of the project, and fail to hold their managers accountable. Along with the cultural traps that many firms embrace, it is easy to see where projects lose money.

Another reason that projects go over budget is due to resource management, and the PMs' failure to delegate. Very often, if the exact resource a PM wants for the project is not available or the project is on a tight timeline, the manager will step in and do the work or have a more expensive resource do the work than what was originally planned. This has the negative consequence of eating up the project margin and affecting the utilization of the less expensive team members.

Mark Zweig, a leading advisor to A&E firms around the world and Founder of ZweigWhite, frankly describes the challenges faced by PMs in the A&E firm:

* *

There's no doubt about it—being a PM is a tough role. As a PM, you may get responsibility for completing a job you didn't start. A lot of bad decisions could have been made that you will have to live with. The fee allocated to do the work may be too low. The client could be impossible to please, yet your job is to please them. You could be stuck with a team of low performers and not have the authority to get rid of them. The job you are assigned to manage may be one of 10 jobs that you are responsible for managing. Your computer system may not support moving the work around the firm to get it completed by those best qualified to do it. There just isn't enough time, money, or manpower to do the job properly. The fact is, project manager is probably the most difficult job in the firm to do well for any number of reasons.

* *

While very true and close to home for many PMs who work for us, it is a sad reality that the job is challenging from the beginning. Add to this the fact that most of them receive little training, especially on the financial side of their project management responsibilities, and you can just imagine the frustration that they feel in trying to be successful.

So what can we do to make their lives easier and at the same time increase the probability that your projects will make a profit? Chapter 5 addresses some of the practical steps you can take to increase your PMs likelihood of success. By focusing on alleviating some of these challenges for our PMs, through improved processes and better systems and training, we have a better chance of increasing our profit margins at the same time.

Unexpected Changes in the Project Scope, Resources, or Other Contingencies

Sometimes things don't always go according to plan, and we are faced with contingencies that drive up our project costs. While there is little we can do to control or exactly predict many of these occurrences, some safeguards and processes can mitigate our losses. We already addressed the practice of budgeting for contingencies. By planning for unexpected situations, we can protect our profit margins from surprises.

While budgeting for contingencies is often common practice for large projects, especially large government or construction projects, many A&E firms do not plan adequately for contingencies on smaller projects and often feel the pain of these unplanned complications in the bank account.

The Association for the Advancement of Cost Engineering, or AACE International, has defined contingency as follows:

• •

An amount added to an estimate to allow for items, conditions, or events for which the state, occurrence, or effect is uncertain and that experience shows will likely result, in aggregate, in additional costs. Typically estimated using statistical analysis or judgment based on past asset or project experience. (Source: http:// en.wikipedia.org/wiki/Cost_contingency)

• •

A useful step often overlooked during the estimating process is to return to previous projects with the same client and review where scope changes, personnel problems, and budget overruns occur. This can lend some insight into how to predict similar circumstances in future projects. Clients tend to have repeat behaviors and issues. By looking at detailed project history for similar projects, especially with the same client, you can better predict potential contingencies that will slow down or throw a wrench into the gears of your project, and budget for them accordingly.

Calculating the Cost of Poor Project Management	
Number of projects worked on per year	40
Average project revenue per year	$250,000
Percentage of projects that go over budget	40%
Percentage amount that projects go over budget	20%
Amount of average budget overrun	$50,000
Total loss due to budget overruns	$800,000
Value of 1% improvement in project profit margin ([40 x $250,000] x .01)	$100,000

Cost of a Long Invoice Cycle and Poor Cash Flow

As the economy has worsened over the last five years, the days sales outstanding for most A&E firms has increased accordingly. Cash flow management has become increasingly important, and if it were not for the extremely low interest rates at the moment, many firms would not be able to survive because of the cost of debt.

The invoice cycle in many firms is longer than it needs to be. This is due to many factors, including poor timesheet management policies and processes, inaccurate timesheets, managers holding up invoices due to concerns about the project, and inefficient billing and collection practices. The reasons that some of these poor management practices are in place refer back to the cultural traps discussed in Chapter 1.

Ted Maziejka, LEED AP and consultant to ZweigWhite, Fayetteville, Ark., explains the importance of focus on cash flow and collections as part of a healthy business management practice:

• •

Cash flow drives the business, and without a continuing rigor at all levels of the organization, the needs of the firm will not be sustainable in the face of challenging economic and financial times. Of paramount importance is to have a clear and concise strategy to effectively manage this effort, and to insure that this strategy aligns with the contractual terms of your client agreements.

All positions in the organization must be working in concert to adhere to the accounts receivable process and procedures: from the accounting staff, the project team, and the senior leadership of the organization.

With the current economic pressures and the challenges facing firms of all sizes, the ability, or rather inability, to effectively collect the outstanding accounts receivable has become a key element to the survival and health of firms.

• •

Cash flow is directly impacted by poor timesheet and billing processes. Weak timesheet policies, or failure to enforce the ones you have, will result in inaccurate invoices, and extended days to get the bills out the door. Both of these problems will cause your client to take longer to pay you, and in some cases, waste multiple billing cycles in justifying and revising invoices for approval. It also has the other undesired effect of causing your client to mistrust your administrative processes, creating potential client dissatisfaction and relationship stress.

While many of these factors can be remedied through improved processes and policies, the firm principals must be committed to change, especially to avoid the trap of not enforcing critical company policies (see Chapter 1). In many cases, the firm may have many of the necessary policies in place in theory but does not enforce them.

You may not have thought about your timesheets as a main cause of poor cash flow. Sometimes this connection is not made by firm executives, and not enough attention is given to timesheets as a way to improve all aspects of the firm. As you will see in Chapter 6, timesheets are critical to a highly performing A&E firm, and many suggestions for improving overall timesheet effectiveness are provided. We will also look at many ideas for improving cash flow through improved collections efforts.

Calculating the Cost of a Long Invoice Cycle	
Annual billing	$10,000,000
Average monthly billing	$833,333
Average accounts receivable (AR)	$2,054,775
Average number days from 15th of month to collection of cash	75
Cost of borrowing (average line of credit interest rate)	6%
Cost per day of borrowing to collect	$1,644 (($10M x 6%) ÷ 365)
Amount required to finance average collection period (75 days)	$123,300 ($1644 x 75)
Amount of savings from reducing collection period to 60 days	$24,660
Average (median) collection period *	73 days

*From *ZweigWhite 2012 Financial Performance Survey.*

Cost of Inefficient and Non-Integrated Systems with Multiple Silos of Redundant Data

As employees go through their daily work, the efficiency of their work environment has a huge impact on their productivity. If I walk into most 50-person firms, I will probably find at least 50 client databases.

Each person has one in their Outlook or e-mail system. Additionally, most firms have a server with marketing and accounting documents, proposals, spreadsheets, an accounting system, and possibly a CRM, and so the number of data sources expands.

The other problem lies in figuring out which database is accurate. With people working harder to get more done these days, it is much easier for mistakes to be made. Having all of these separate systems and redundant data sources makes it difficult to maintain accurate data and presents challenges for employees to find what they are looking for quickly. There is a huge cost associated with the daily waste of employee's time spent looking for information and entering data into multiple systems.

The truth is that technology is much cheaper than people. By investing in technology systems to integrate as many separate data sources and processes as possible, you can free up your employees to be more productive and potentially extend the amount of time that you will have to hire additional administrative resources as you grow.

If your goal is to grow, having antiquated and non-integrated systems will create bottlenecks that will limit your growth, and give your more automated competitors a direct advantage.

There are many areas where cost savings can be recognized by integrating systems and data. We have already gone over several of these, such as improved proposal processes, and faster and more accurate billing. A good system will impact every one of the areas where you can find lost dollars in your business. We will look more specifically at what to look for and how to select and implement these systems in Chapter 9.

Calculating the Cost of Wasted Time Due to Redundant Data Entry and Looking for Accurate Information	
Number of full-time employees	75
Average labor cost burdened at DPE	$40/hour
Number of hours wasted per week	2 hours per week
Annual cost of wasted time	$300,000
Amount saved per year in 1 hour per week with integrated systems (75 × [$40/hour × 2 hours × 50 weeks])	$150,000

Cost of Losing a (Good) Client

As discussed earlier in this chapter, each firm has its good and bad clients, and some in between. Finding the right balance in working with both new and existing clients can be a challenge. Based on the traps identified in Chapter 1, it is difficult to find clients who value our work and do not take advantage of our focus on quality over profit. We must balance our focus between keeping our existing clients, and hunting for new clients.

I am sure we can all agree that keeping our "A" clients should be a priority, but clients sometimes do not want to work with us again. Most of us do not want to consider this possibility. However, in today's competitive environment, it has become harder to win and maintain client relationships and satisfaction.

The truth is that it is much more costly to find and win a new client than it is to retain an existing one. In her popular blog, Laura Lake provides the following statistics that are hard to ignore:

Before you spend your time and money going after new customers and clients you do not currently have a relationship with, consider the following statistics:

- *Repeat customers spend 33% more than new customers.*
- *Referrals among repeat customers are 107% greater than non-customers.*

- *It costs six times more to sell something to a prospect than to sell that same thing to a customer."*

(Source: http://marketing.about.com/od/relationshipmarketing/a/crmstrategy.htm)

Specific measures can be put in place to improve client communication, assess client satisfaction, and request client feedback. Unless this is your firm's focus, you will find this difficult to accomplish.

While this may not seem like a financial management issue, it has such an impact on the overall profitability of your firm that I felt compelled to include it. In looking at this from the opposite point of view, how much can you save by retaining all of your existing "A" clients and ensuring that they will return when they are looking for help on their new projects?

One area to focus on is how good your employees are keeping your promises to your clients. For clients, the little things like calling back when you say you will, are very important and affect a client's opinion of your firm's quality of service. The truth is that most principals do not know how well they are keeping their promises to their clients. If this is true for you, then this is likely to be the one thing that keeps you up at night.

Many firms spend so much time looking for new business and new clients, they fail to appreciate the great investment they already have in existing clients. Viewing your client base as an asset that needs to be protected is the first step to creating a culture focused on client retention.

Client feedback can also help you understand what services your clients truly value and assess what direction to take the firm during your strategic planning processes. Make it a priority to improve and document client correspondence, and ensure your staff is keeping promises to clients. This can add a tremendous amount to the bottom line and help you discover the lost dollars in your existing client base.

Mike Phillips, President of Phillips Architecture in Atlanta, Ga., provides the following insight into how to determine your firm's performance in the eyes of your client:

. . . .

Since [A&E] firms' success is dependent upon how they deliver services to their clients, the most strategic way to measure results is a combination of financial metrics (project profit) and client perceptions (feedback). Adding a system for quantifying client satisfaction with detailed metrics allows a firm to fine-tune their process of delivery to better meet each client's unique expectations. Because we measure to improve, measuring from the client's perspective most quickly creates the understanding of how to increase your firm's value, and your client's loyalty.

. . . .

Calculating the Cost of Losing One Client	
Average number of clients per year	25
Average revenue per client	$400,000
Cost of losing one client	$400,000 per year
Amount saved by retaining one additional client per year	$400,000

Summary

There are nine areas of your business where revenues gets lost or money is wasted because of ineffective processes, poorly trained people, and inefficient or missing systems. We measured the cost of not optimizing each of these nine areas for a firm with $10 million in annual revenue. The nine areas where money can get lost in the A&E firm include:

1. **Cost of lost opportunities:** Good decisions must be made in deciding which projects to pursue. The lack of a defined sales process and effective tracking of open opportunities will cause a lower-than-desired win rate.

2. **Lost revenue from a cumbersome proposal process:** Writing and assembling quality targeted proposals is critical to success in winning work. If data is not managed effectively

and proposals are cumbersome to produce, time is wasted, and win rates can suffer.

3. **Lost revenue due to a flawed estimating process:** The estimate is critical to ensuring a project will be profitable, yet many firms do not have control over the estimating process. The cost of this inconsistency is poor estimates, which result in poor contracts and budgets, as well as projects that are harder to manage.

4. **Lost revenue and excess costs from scope creep or extra services not recovered:** Scope creep is caused by many factors including the 10 traps unveiled in Chapter 1. Control of extra services is critical to ensuring that the firm is not giving away services for free.

5. **Cost of low utilization and poor resource scheduling:** The ineffective scheduling of employees and forecasting labor requirements into the future will make hiring decisions tough and will cause poor utilization rates.

6. **Cost of poor project management:** Poor project management is caused by flawed budgeting processes, using resources that are not planned, and not planning for changes affecting the cost of the project. These problems can have a substantial effect on project profit margins.

7. **Cost of a long invoice cycle and poor cash flow:** Poor cash flow causes the firm to have to borrow money and can strain client relationships.

8. **Cost of inefficient and non-integrated systems with multiple silos of redundant data:** Having data in many different places is inefficient and causes employees' time to be wasted in redundant data entry and manual tasks.

9. **Cost of losing a client:** While we know that all clients are not good clients, we do not want to lose the profitable clients. Establishing processes to ensure outstanding client relationships is critical to firm success, and is ultimately less expensive than finding new clients.

BEST
PRACTICES
FOR IMPROVING
PROFITABILITY

STRATEGIC PLANNING

"If you don't know where you are going, you are certain to end up somewhere else."

~YOGI BERRA

For anyone that has been in business for any length of time you have probably engaged in a business planning process. Many firms do this once a year, while some smaller firms may have never formally created a strategic business plan.

It is a very beneficial process to go through, especially for a firm that is growing or wants to grow. A good strategic business plan will start with the mission, vision, and values (MVV) of the firm leaders. This is important in how it shapes your conversations with your employees, and creates the groundwork for setting behavior expectations and accountability measures. It also gives your clients a way to understand your firm's philosophy about how you do business, treat clients and employees, and your vision for the future.

You may want to hire a business coach that has experience working specifically with A&E firms to define and document your MMV, and to define your strategic plan. There are many excellent advisors out there, including several who are quoted in this book. In addition, numerous organizations such as ZweigWhite and PSMJ provide guidance in helping A&E firms develop a strategic approach and plan.

In addition to the MVV, you should do an analysis of your strengths, weaknesses, opportunities, and threats (SWOT). Strengths and weaknesses are internal to your firm, and opportunities and threats are conditions that exist outside of your firm that provide a path to follow or situations to avoid. Figure 1 is an example of a standard format for doing a SWOT analysis.

Strengths	Weaknesses
What are your strengths? What are your competitive strengths? What are your perceived versus real strengths? What other internal factors help you excel?	What are your weaknesses? What are your competitive weaknesses? What are your perceived versus real weaknesses? What other internal factors will hinder you?
Opportunities	Threats
What current opportunities exist? What future opportunities may exist? What competitive opportunities exist? What other external factors may help you excel?	What current threats do you face? What future threats do you face? What competitive threats do you face? What other external factors will put you at risk?

Figure 1: SWOT Analysis

(Source: http://www.sales-and-marketing-for-you.com/support-files/swot-analysis.pdf)

While I have been doing business planning for many years for my own businesses, I am certainly not an expert in it and have not consulted with clients as my area of expertise. I found a great process for this in,

Mastering the Rockefeller Habits, by Verne Harnish. This is an excellent resource for any business owner who is trying to put together an effective and relatively simple strategic plan. Harnish has an excellent strategic plan that is extremely useful for getting all of your strategic planning documented in one place. The one-page plan includes the MMV, SWOT analysis, a way to document your long- and short-term goals, and the key performance indicators (KPIs) that you will use to assess your business performance.

Harnish also describes a regular rhythm of meetings with key staff that help keep the performance goals front, center, and on track. Each of these goals and KPIs flow down to your staff to form employees' individual performance measurement system. Because you are looking at your numbers and each person's performance on a regular basis, there are not as many surprises during the annual review process. Go to www. Gazelles.com to download several documents that can help you execute all of the strategic planning advice described in the book.

Part of your strategic planning process should also include talking with your employees and best clients to understand how things are going from their perspective. This can provide very valuable feedback and help you determine which way you want to take the company. It is great to realize that you don't have to do this alone! Your valued clients and employees will actually be excited to help you determine your future course. Their individual and unique outlook and ideas will provide a well-needed shot of reality, as well as some potentially new ideas you had not previously considered. It will also have the side benefit of helping you get buy-in to your new strategy and planned success.

Business strategy can have a huge impact on your firm's profitability. By targeting the right markets, and developing a well-thought-out marketing plan and specific goals, you can identify your firm's unique offering and position yourself among your competitors.

Lee W. Frederiksen, Ph.D., author and managing partner of Hinge Marketing in Herndon, Va., provides some excellent advice for A&E firms trying to figure out how to determine where to focus their marketing efforts:

* *

One of the most common strategic mistakes that we see A&E firms make is to target their marketing and business development efforts too broadly. There is a very human reaction to try to offer as many services as possible to as broad an array of potential clients as possible. The emotion is fear of turning away any potential client.

In reality the research we have conducted on hundreds of firms clearly shows the opposite. The more narrowly you target your services and the deeper you understand your industry, the greater the likelihood that you will grow faster and remain profitable. Clients prefer a specialist who understands their industry in depth and has solved similar problems many times before. **With the rise of the internet, geographic proximity is becoming less important and specific knowledge and experience is becoming more important.** *Of course there are limits. You must target a niche where there is sufficient business and your expertise must be real. But narrow and deep almost always wins over broad and shallow.*

* *

Strategic planning should include the following elements, which will enable you to develop an annual marketing budget:

- What is our longer-term end goal for ownership transition? Hobson Hogan of ZweigWhite advises that it takes at least 10 years to plan, implement and fund a transition plan for the younger generation of leadership. If your goal is to have your younger managers eventually become principals and buy out the current owner's interest, you had better allow 10 years to get everything in place or find an alternative transition plan.

- How much revenue and profit do we want, or expect, to generate in both long- and short-term increments (5 years, 2 years, 1 year, etc.)?

- Who is our target client? This includes looking at the industries, geographies, size, demographics, and type of client (public versus private) that you will pursue.
- What services will we offer?
- What types of contracts will we pursue?
- How many people will we need to hire (or let go) to achieve our expected goals? This is also driven by determining if you have the right talent to achieve the goals you are setting.
- What will be our primary methods of acquiring new clients?
- Should we consider an acquisition as part of a strategic initiative to move into a new market, or offer some new or expanded services?

All of these questions, and possibly many others, are critical to getting and keeping your team focused in the right direction, and achieving your firm's strategic goals and objectives.

Developing this roadmap for your company's future is the critical first step to ensuring that you have set the course and are traveling in the right direction. The next step is to build the context with which you will measure your company's performance over the next year.

COMPANY-WIDE FINANCIAL BUDGETING

Once you have determined your business strategy, and long- and short-term goals, you should create your annual budget. This is a critical aspect of good financial management, and serves two purposes.

First it forces you to take a deeper look at your numbers each year and evaluate if your current financial management processes are working. The company budgeting process can help you answer some important financial management questions to help determine critical business decisions that need to be made. The key questions that the budgeting process can help you evaluate and answer include:

- Are the accounts set up so that you can really assess the performance of the company?
- Does the bottom-up budget align with the top-down budget?
- How do we compare to industry averages?
- How will we measure the success of individual business units?
- What do we need to do to control overhead?
- How much are we going to spend on marketing and where?

BUDGETING BEST PRACTICES

The beginning of each fiscal year is the best time to make changes to your chart of accounts, financial statement formats, and your budget line items. It can also help you break out information for taxes as well, so I recommend that you consult with your tax accountants to see if they have any recommendations on breaking out expenses further.

In addition to looking backward, creating a detailed financial budget can be a great exercise for your managers. It helps them remain more focused on their role in the financial success of the firm. Having a strong budget that is based on past experience and the financial goals of the company will help you assess on a monthly basis whether you are on track or need to make adjustments.

Your budget will comprise estimating annual revenue and then determining the estimated expenses that will be required to support the company's forecasted performance level. I recommend doing both bottom-up and top-down revenue budgeting for your company and comparing the two to see if your budget numbers are on target.

Top-down budgeting is where you set financial targets and create the budget based on revenue figures that you ideally want or need to hit in order to cover costs and make the profit that management expects.

Bottom-up budgeting is where you look at your individual resources and determine how much revenue they can realistically generate based on their expected utilization and performance, as well as the company backlog and pipeline.

After you have calculated your expected revenue both ways, you can compare the two to see if they are close. If they are close, then you are probably on target with your forecasts and should have a reasonable revenue budget. If the numbers are far apart from each other, then you will need to re-examine the details to understand how to make necessary adjustments. In most cases, the top-down budget will be higher than the bottom-up one. Because the bottom-up budget is developed from a more detailed analysis of your actual resources and project backlog, it is usually more accurate and better represents your potential next annual performance.

For your expenses, your past history can be extremely valuable in assessing future expenditure requirements. I recommend that you look at prior years' numbers as a percentage of revenues, as well as break down each general ledger (GL) account based on expected changes. For example, you should estimate labor expenses based on actual payroll amounts, taking into account projected staffing changes. Marketing expenses should be backed up with a detailed marketing budget.

If you have multiple profit centers or departments, your budget should be further broken down. Based on how your financial accounts are structured, you can manage these organizational units separately, which requires allocating all revenue and expenses to the individual business units. An easier method is to have separate GL accounts for just the revenue in your accounting system and assigning projects to them accordingly. It really depends on whether you want to see a separate profit and loss (P&L) statement for each business unit, or just the revenue broken out. Certainly the latter is much easier to manage and requires a lot less scrutiny and possibly subjective determination of how to allocate the expenses.

The majority of A&E firms manage their books on both a cash and accrual basis. The accrual basis is a much more accurate way to evaluate firm performance. It measures how the company performed during each period using generally accepted accounting principles (GAAP). This requires that you take into account the revenue billed for the period, as well as the expenses incurred (committed but not paid for yet).

The cash basis of accounting only looks at the cash that came in and went out. While it is very useful for understanding cash flow and tax liabilities, it does not really give a good picture of how the company performed during the period. The cash basis balance sheet also does not include AR and accounts payable (AP) that give a better picture of the health of the firm at a given point in time. We will delve further into the pros and cons of these two accounting methods in Chapter 7.

Once the budget has been completed, you should be comparing it to your actual revenue and expenses on a monthly basis. The difference between the budget and actual figures is called the *variance.* Reporting should be developed and made available to firm principals showing trend analysis and the variances as a percentage of revenue. You should also be evaluating many other specific metrics each month. Chapter 7 goes over the KPIs you can build into your monthly reporting process to help you analyze firm performance.

Summary

- Strategic planning is essential to consistent achievement of performance and growth goals. A strategic plan starts with the Mission, Vision and Values of the firm. From there, analysis should be done to assess the firm's strengths and weaknesses (internal), and opportunities and threats (external).
- Effective strategic plans should include the following components:
    - o Ownership transition goals
    - o Revenue targets
    - o Definition of the ideal client and target audience for marketing
    - o Detailed list of services to be provided
    - o The types of contracts to be pursued
    - o Staffing and hiring expectations

    o   Marketing strategy and detailed plan or budget

    o   Potential for acquisitions as a way to achieve strategic goals

- Company-wide budgets should be prepared annually to project the revenue and expenses for the coming year, based on expected growth plans. Both a top-down and bottom-up version should be prepared and then evaluated against one another to see where adjustments need to be made.

- Financial reports should be produced on a monthly basis, which compare the budget numbers to actual financial results for the period and year-to-date.

CHAPTER 4

MARKETING, BUSINESS DEVELOPMENT, AND PROPOSALS

Marketing is not an event, but a process ... It has a beginning, a middle, but never an end, for it is a process. You improve it, perfect it, change it, even pause it. But you never stop it completely."

~JAY CONRAD LEVINSON, AUTHOR, GUERILLA MARKETING

For many A&E firms, marketing and business development are elusive. This is because many owners start their businesses with one or two clients that supply most of the work. They often only have to start going after business as their firms grow and competition becomes more of a factor.

What many A&E firms call marketing is actually business development and proposal preparation. True marketing is a many-to-one function of the business, involving branding the company to establish its reputation, advertising, public relations, trade shows, pursuing awards, and inbound marketing (online lead generation), including blogging, social media, and creating marketing campaigns to generate potential leads.

Business development (BD) is the specific function of developing industry and client relationships, looking for opportunities, and hunting for business. This is distinguished from proposal preparation, which is the specific pursuit of a single project, usually as part of a team.

Many smaller A&E firms do very little marketing as defined above, and have most of their efforts in BD and proposal preparation. While this is often referred to as marketing by many in the industry, there is really a huge difference between the many-to-one approach and the pursuit of a project or response to a request for proposal (RFP).

The Society for Marketing Professional Services (SMPS) is an excellent resource for principals, and marketing/BD professionals. With a national and local chapter presence, your marketing/BD staff can have access to valuable resources to help them fine-tune the company's focus, learn best practices for all aspects of promoting the firm's unique specialties, and network with others in the local area for potential teaming opportunities. Ron Worth, CAE, FSMPS, CPSM, and CEO of SMPS describes the organization and its mission as follows:

* *

The Society for Marketing Professional Services (SMPS) was created in 1973 by a small group of professional services firm leaders who recognized the need to sharpen skills, pool resources, and work together to create business opportunities. Today, the association represents a dynamic network of 6,000 marketing and business development professionals from architectural, engineering, planning, interior design, construction, and specialty consulting firms located throughout the United States, Canada, and the United Kingdom. The Society and its 50+ chapters benefit from the support of 3,800 design and building firms, encompassing 80% of the Engineering News—Record Top 500 Design Firms and Top 400 Contractors.

SMPS offers members networking and leadership opportunities, a national marketing conference, workshops and webinars, a certification program (Certified Professional Services Marketer), awards programs, publications, and educational resources to highlight best practices and the latest trends in business development and marketing in the design and construction industries. In addition, SMPS' active chapters offer local and regional networking and professional development to members and their firms. To learn more, visit www.smps.org.

THE CHANGING WORLD OF MARKETING

Marketing has changed dramatically in the last few years, as a result of technology and primarily social media. The way people interact, share information, and make announcements is completely different than it was just two years ago. I used to think I was a pretty savvy marketer, only to discover that I had to learn everything all over again.

An example of this is how the Washington Redskins NFL team announced on Twitter that its star rookie quarterback, Robert Griffin, III, (@RGIII on Twitter) would be inactive in the game against the Cleveland Browns in October 2012 because of a knee injury. Just a couple years ago, this would have been announced on television or the radio. Now Twitter is one of the primary places to get news, find out what is trending in the world, and join in on the conversations:

Figure 2:

If your marketing team has not embraced the new world of inbound marketing, then your firm is not on the competitive side of the new marketing revolution. Building a presence in social media, in order to be able to take advantage of the Internet as a marketing tool, is a slow process that takes a persistent, steady investment of time and resources. The benefits are realized gradually over time, as your firm, and the thought leaders within your firm, start to gain traction and build a platform.

Mark Amtower, social media expert, founder of Government Market Master, and author, gives the following advice to those who are not yet taking advantage of social media as a channel for acquiring new business:

• •

If for any reason you are hesitant about using social networking, consider this: As of 11/30/12, LinkedIn has more than 1.5 million groups, precisely 1,500,638. This number grows hourly. Further, as of the same date, there are 498 groups that deal with [AEC]. These are self-identified enclaves of professionals in your niche, several with thousands of members. Add to this the various facilities and business owner groups. Both IFMA and BOMA have robust presences on LinkedIn. Among these, you can find people and companies that require services like yours, and companies that may complement what you do, not to mention potential employees. If you opt to remain outside of the social networking "scene," understand that some of your competitors will profit from your absence.

• •

Just having a Web site is not enough these days. In order to get noticed on the Internet, you have to have content that people want to read, and which drives the site engines to your site. One of your goals should be to increase search engine optimization (SEO), or the number of times your site comes up in search results in Google, Bing, Yahoo, or any of the other popular search engines. Content is what drives traffic, and it must be constantly generated and accessed in order to attract interested parties.

A blog is the primary way to get content online for potential followers, clients, and teaming partners to find you. It is a collection of articles, videos, and other targeted information available on a Web site and optimized to be found by people searching certain keywords. The subject matter in the blog must be interesting, relevant, and optimized for SEO in order to attract attention. Utilizing social media platforms such as LinkedIn, Twitter, and Facebook are the primary ways to get your content in front of potential followers (i.e., potential clients and partners). Having valuable content that your target audience wants to read will enable you to build a following on these social media sites and exponentially multiply your potential viewing audience across the world. In particular, Twitter will allow you to reach millions of people you don't know but who might have an interest in the topics you are writing about.

The process of producing the content for a blog is hard work and time-consuming. I am writing about one to two articles a week for our blog (http://AcuityBusiness.com/blog), as well as six articles a year for the *Zweig Letter*. Over time, it will start paying off, as you build followers and attract visitors to your site. You should have landing pages and calls to action (CTAs) on your site, which will allow you to convert a visitor to a lead. A CTA is a button on the site that allows visitors to sign up to receive your blog by e-mail, download a whitepaper, subscribe to your e-mail, or ask to be contacted.

I highly recommend using a system such as Hubspot or WordPress to manage your blog and analyze the traffic to your site. These systems can help you quickly understand who is visiting your site and which pages or articles are the most popular, as well as provide a process for tracking leads that have been converted.

While A&E firms have been slow to adopt social media as an authentic marketing vehicle, Brenda S. Stoltz, CEO of Ariad Partners, and marketing advisor to my company, Acuity Business Solutions, has the following insight about social media for A&E firms today:

. . . .

The easiest way I've found for people to understand the three primary social media platforms is to say that Twitter is like a bar, Facebook is like your living room, and LinkedIn is your local Chamber of Commerce. You never know who you'll meet on Twitter,[and] there can be a great deal of serendipity. You can be in and out of Twitter in a couple of minutes— have a beer and head out. With Facebook, there is more work. You have to put more time into creating opportunities for engagement on Facebook. You wouldn't invite someone into your home for a visit without a chair to sit down on, or a cup of coffee or tea for them, the same is true or Facebook. You must make sure you have content and opportunities for engagement. And finally, LinkedIn is all about business—all business all the time. Behave with decorum. Don't waste anyone's time and always be professional.

When I hear from executives that they don't see the value or ROI of social media, I ask one question, "Do you want to be where your customers are?" Of course, the answer is always yes. Well, your customers are on social media. At that point, it usually takes a quick search on Twitter, Facebook, and LinkedIn to show them which platform (if not all) their clients are using, their industry associations, thought leaders, and yes, sometimes (oftentimes) we even find their company name being mentioned. It is a shock, but they finally get it.

. . . .

As Brenda implies, implementing a social media marketing strategy is not only, but also essential in today's world. Because it takes so long to build a library of valuable content and a "tribe" of followers on social media, the work you do today may not pay off for a year or more. I advise you to get ahead of the curve now, and beat your competitors to the punch by diving into social media and inbound marketing as a way to build your company's brand and platform.

THE GO/NO-GO DECISION PROCESS

Earlier we looked at the importance of the Go/No-Go decision in an A&E firm's evaluation of whether to pursue an opportunity. Certainly the value of this process is to save the company money by limiting bad choices and increasing the firm's overall win rate. And as Bob Gillcrist pointed out, it can also save the company down the road by eliminating the potential for the firm to win a project for which it is not suited or will lose money.

In order to put together an effective Go/No-Go process, you need to determine the criteria that will help you assess both the likelihood of winning the project, and how successful your firm can be at executing on it. While there are many common approaches to this, here are a few high-level questions that need to be answered:

- What is our relationship with the potential client?
- Who are our competitors and what is their relationship with the potential client?
- Have we done business with them before?
- Is our cost competitive?
- Is this our "sweet spot" considering our project history or outside of our normal skill set?
- Do we have the qualifications they are looking for in the RFP?
- Can we make money on the project?
- What will it take to respond to the RFP?
- Can we put together a successful team?
- Can we do the work?
- Will we be successful at this project?

You should assign a point system for each criterion, in order to calculate a final score. Here is an example of a Go/No-Go grid that you can build into your client relationship management (CRM) system:

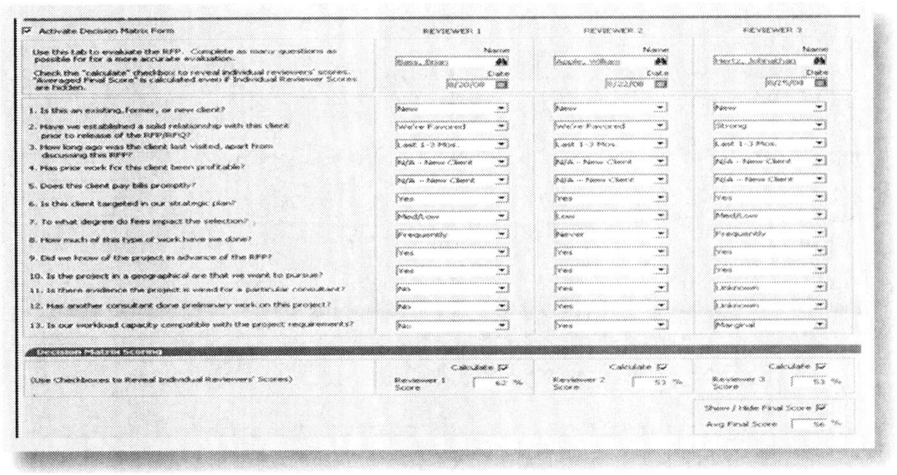

Figure 3:

**(Courtesy of Brian Bass, Sr. CRM Consultant for
Acuity Business Solutions, Deltek Partner.)**

A detailed Go/No-Go strategy will analyze the potential of the opportunity in four areas: client relationship, project criteria, competitor comparison, and team information. Obviously the more homework you do, the more accurate the results of the analysis will be. This may not be a viable process for small projects, and you should be investing the appropriate amount of time evaluating and responding to each opportunity that really makes sense, depending on its relative size and potential profitability.

Another important aspect of the Go/No-Go process is to make sure the company is pursuing the "right" projects. Tim Klabunde, Marketing Director of Timmons Group, Richmond, Va. describes the need for the Go/No-Go process to be aligned with the company's overall business strategy:

● ●

A Go/No-Go decision-making process that is not tied to long-term objectives will almost always cause vital company resources to drain away from where you want your company to go, and towards where you have always been.

● ●

In looking at your win rate, as well as the level of success you are experiencing in picking the right projects to bid on and accept, there is an inherent ROI you are realizing from your investment in the proposals on which you are bidding. By understanding the cost of each lost proposal and the criteria used to pick the winning team, you can start to define a Go/No-Go strategy and criteria that are more appropriately rewarding your efforts.

The expression *opportunity cost* is particularly relevant in this analysis, because for each opportunity you pursue, you are foregoing the ability to bid on another, or invest the money in other areas of the business that will provide a greater payoff.

One way to evaluate this is by tracking the costs of opportunity pursuits in a promotional project in your system. By having everyone involved with the opportunity and preparation of the proposal track their time and expenses on a specific project, you can start to better analyze the effectiveness of the decisions being made, as well as the resulting ROI from the ones you have won.

OPPORTUNITY MANAGEMENT

Having a streamlined and comprehensive opportunity management process is a critical factor in achieving competitive success, as evidenced by a high win rate. This is managed in firms in many different ways, from spreadsheets to basic or very sophisticated CRM systems. A good CRM system, along with a detailed and well-executed opportunity management process, will ensure that your firm can stay competitive and continue to bring in new work. It will also help management assess the success of its marketing efforts and provide better visibility into future project revenues.

Depending on how your opportunity management process is established and who in your firm works on pursuing new business, you may have information in many different places. A CRM system can provide a valuable tool for consolidating all the data about your projects, people, qualifications, and opportunity history in one place. Consider a few of the many benefits of managing your marketing and business development data in a CRM system:

- Gives management the ability to report on pipeline for future backlog and revenue forecasting.

- Consolidates tracking of client and prospect contact information to improve the accuracy of the firm records and eliminate the need for multiple databases.

- Provides visibility into short- and longer-term staffing requirements for better hiring and staffing decisions.

- Helps to ensure that important proposal deadlines are not missed.

- Allows recording of completed and pending activities in order to store historical information, as well as provide to-do reminders for deadlines and effective follow-up.

- Enables collaboration and sharing of resume and project data between multiple teams and offices on the opportunity capture process.

- Allows you to calculate your win rates to assess marketing success.

- Provides a place to store competitor information.

- Facilitates the Go/No- Go analysis, and in many cases, can fully automate it using workflows and built-in calculations.

- Stores team and project information for each opportunity.

- In some systems, the opportunity data can feed directly into a proposal or SF 330, as well as be used to set up the project when won.

- If integrated with accounting, can streamline the project setup process, and make it easier to gain access to critical project and employee qualifications data for proposal preparation.

- Can help systemize the sales process for better consistency, accuracy, and approvals.

- Reduces risk from turnover of staff involved in pursuing opportunities.

A good CRM system will allow you to track all of the leads and opportunities that your firm is pursuing. It should also address all of the issues listed above, making your sales team's job easier and better enabling you to hold the responsible parties accountable for achieving the firm's new business goals.

If your firm is not using a CRM system, then you are potentially allowing valuable firm data and project opportunities to slip through the cracks. You are also not operating your firm with the best information possible. An industry-specific and project-based CRM can make it even easier to adapt the system to your process, and evaluate the effectiveness of your proposal efforts.

PROPOSALS

Logically, if we are improving our estimating and opportunity systems and processes, we should see better results. This is true to a greater extent if an integrated system is in place. Getting our act together in the firm's marketing and estimating should also make it easier to prepare quality proposals. It is definitely easier than it was 20 years ago, when computer systems were just starting to be developed for A&E marketing and proposal development.

In 1992, I was working with a piece of software called *A/E Marketing Manager*. In looking at proposal practices at the time, especially with the SF 254/255, there was little in the way of automation, and many firms were using practices that today would be viewed as bizarre. In order to fit more information in the little boxes on the form, I recall one firm had enlarged the form on a copier, entered the information on a typewriter, and shrunk the form back down. This was an archaic process!

We are all grateful we don't have to do these types of cutting and pasting tricks today, as computer systems have made it much easier to format proposal templates and adjust the data to fit. However, I still see many firms that have unsophisticated processes, disparate data sources, and antiquated systems. This leads to a less than optimal work environment for your proposal contributors. Making their lives easier should result in improved win rates.

The other major issue in effective proposal development is how and where the data is managed and retrieved during the data-gathering portion of the proposal effort. If your marketing team cannot easily locate the key information about your projects and staff qualifications, or have the ability to produce targeted resumes and project descriptions, the process will slow down to a crawl.

Nancy Usrey, FSMPS, CPSM, author and leading expert in A&E firm proposal development, spoke at a recent webinar for my firm called, *SF 330: Secrets to Getting Shortlisted.* In this webinar, she offered the following advice for winning success on government proposal submissions: "Prove your experience with focused project descriptions and select those projects that best illustrate the *team's* qualifications for *this* contract."

Certainly this is only possible with excellent data management that easily enables your proposal team to answer such questions as:

- What projects has the team or company done that are similar?
- Who worked on them, and what were their roles?
- What did we do on these projects?
- What were the budget and construction costs?
- When did the project start and end?

If your firm has a marketing database with all of this data, then your proposal team will have a competitive advantage over those firms whose staff has to walk around the office and try to extract information from people's heads, or cut and paste from old proposals.

The key to looking for the lost dollars in your proposal process is to understand your win rates and how you might be able to improve them. You can also look at the number of proposals you are submitting and determine if a better process would enable you to submit more.

Certainly the major difference between 1992 and 2012 in the tools we have available for preparing proposals is in technology. Without adequate systems for managing marketing data and streamlining the proposal assembly process, both quantity and quality are affected.

I find that bottlenecks and deficiencies in the typical proposal process occur in a few primary areas:

- Access to key qualifications and project data housed in the accounting system or other databases
- Cumbersome process for assembling the data in the proposals
- Locating the right boilerplate documents for customization in a proposal
- Management of templates for résumés, project detail sheets, etc.
- Consistency in the proposal formats between teams
- Lack of targeting due to using a few standard templates for every proposal
- Lack of collaboration and sharing of data between multiple teams or offices
- Keeping all of the data updated

By addressing the specific bottlenecks your company is experiencing in the proposal process, you can see remarkable results in both the quality and quantity of proposals you are able to submit. As most firms use a certain amount of the PM's time in proposal development, this will also free up time for your most valuable resources, and potentially enable them to be more billable.

A good proposal management system will be completely integrated with your CRM, and optimally, your accounting system. This will help alleviate some of the friction between marketing and accounting in sharing data. Additionally, the system will allow flexible template formatting and ability to control the templates the staff uses.

Summary

We have examined many of the processes and systems that can help your firm win work. Best practices have been identified for determining which projects your firm should pursue and how opportunities are managed, and for preparing winning proposals.

The following are the main points that contribute to improving the firm's win rate:

- Marketing is changing rapidly because of technology and the growth of social media. Your firm should constantly investigate how you can improve your brand image online and how leads are found through inbound marketing.

- A good Go/No-Go process includes a consistent evaluation that ensures that the firm has a good chance of winning the project *and* that the project is going to be a successful one for the firm. This includes looking at many factors including your firm's relationship with the client, your qualifications, whether the firm can make money on the project, and your ability to do a quality job and meet the client's expectations.

- To ensure a high win rate, your firm must manage opportunities as efficiently as possible. The best way to do this is to utilize a CRM system, and develop streamlined processes for managing opportunity data, tracking deadlines, and following up appropriately.

- Win rates are also affected by the efficiency of your proposal processes. The more you can enable your proposal teams in the tracking of qualifications data and streamline the development and assembly process, the more successfully and quickly they will be able to deliver high-quality, targeted proposals with less effort.

CHAPTER 5

OPERATIONS

"By working faithfully eight hours a day you may eventually get to be boss and work twelve hours a day."

~ROBERT FROST

Once you have won projects and begin to execute on them, there are challenges in being able to successfully deliver the project, while also controlling project costs. This is where PMs are crucial to the success of your firm and need special attention on how they are trained, managed, and held accountable.

Most managers are responsible for managing projects and people. Some of your employees may work with different managers on multiple projects, or they may work on one project team for a single manager. In either case, the way that the communication, scheduling, and management of their work product is conducted will affect the profitability of your projects.

PROCESSES AND PROCEDURES

There are three primary ways to affect the successful operations of your firm—using people, processes, and technology. Processes enable people and technology to work together, and help them interact and utilize technology more effectively. As a firm grows, it is essential to evaluate and put processes in place for every area of the business, as well as every interaction between people and technology. Without these processes, there will be inconsistent behavior leading to vastly erratic results.

One example is the lack of controlled templates in the project estimating process that we identified above. Without specific processes that control the estimating functions, estimates and proposals could be sent out the door with incorrect rates, inconsistent language and project work breakdown structure, and improper assumptions. Without adequate timesheet approval processes, time could be billed to the wrong projects, which has an adverse effect on cash flow and eventually breaks down client relationships. Looking at these two examples, it is obvious that we need better processes and controls, but how do we evaluate and develop them?

The key to assessing and building successful processes is to evaluate each process, and map out how it affects everyone on the team and flows through your systems. This can be documented both as a flowchart that shows the major communication and decision points, as well as written out in outline form.

I recommend for each major area of process improvement, you assemble a committee to look at the process in more detail and develop the structure of the new recommended process. It is preferable to work on only one major process at a time, as changes made to one process will very often affect the steps required in another process.

The first step is to interview as many people as possible who are affected by the process. This should include employees, clients, subcontractors, vendors, and any other service providers the process affects (e.g., outside accountants, payroll company, or other business advisors). By talking directly to your staff, and others outside the firm who are affected by the process, you will start to get excellent suggestions for process improvement. Your employees and clients will be very grateful for the opportunity to contribute and may come up with some ideas that previously had not been considered.

You should also look at how the current process is driven through your systems. I can't tell you how many times I have seen a process that was established in one system because of limitations in that system, and carried over to a new system out of a desire to minimize change or lack of knowledge about why the process was established. If you are using a system written and developed in the last 10 years, you may find that there are specific features designed to improve a process of which you are not even aware. By consulting with outside system experts, you may find that there are major improvements that can be made just by using the system differently!

One other suggestion is to talk with peers in your industry. They can provide excellent ideas you may not have considered yet. Many others have solved the same problems already and can help you find a shortcut to improving and streamlining your processes.

After the process committee evaluates the current process, interviews internal and external resources, and evaluates the impact on the firm's systems, they may be ready to recommend a new process. At this time, the new recommended process should be drawn out into a new flowchart. At this point it usually helps to run a specific opportunity, project, or some accounting actions through the new process, and to conduct a preliminary test on it. In addition to drawing out the process, you should look at who it will affect, whether changes will need to be made to your systems—both manual and computerized—and determine how you will document, train staff, and roll out the new process.

If you are using an integrated CRM, accounting, and project management system, you will want to investigate how the process affects the various parts of the system and whether the system can be modified to automate the process further. Extensive cost savings can be recognized from automation improvements. These cost savings are a result of greater efficiencies, improved results, and overall project execution improvement.

Timesheet policies and processes are an excellent place to start and can have a huge impact on many facets of the business. By revamping your timesheet entry and compliance, you can dramatically improve timesheet accuracy, recover more extra services, ensure government contract compliance, shorten the billing cycle, improve cash flow, and increase efficiency for managers, accounting staff, and employees. Without these kinds of processes, all of these critical areas of the business suffer, and money is being wasted across the organization:

Simplified Timesheet Entry Policies and Procedures

1. Employees are assigned to a supervisor or group for approval responsibility.

2. Timesheets are to be entered daily by all employees at the end of the day.

3. If an employee does not know what project, phase, or task to charge to, he or she should ask a supervisor or PM.

4. If a project, phase or task is inactive, the employee should ask the PM what to charge their time to.

5. If an employee is working on a task that a client asked him or her to do, which is believed to be outside the scope of the contract, he or she should check with the PM for the correct code to charge to (see policies for handling of extra services).

6. If an employee is out on leave, he or she can complete a timesheet on the day of return.

7. Timesheets are submitted at the end of each pay period.

8. Supervisors approve timesheets and determine if changes need to be made. Supervisors alert employees to any requested changes, and only employees can change their own timesheets.

9. Accounting administrator reviews timesheets for overtime, paid time off (PTO), and other irregularities, and rejects timesheet if the employee needs to change it.

10. Employee makes corrections to timesheets and resubmits.

11. Supervisor approves timesheet.

12. Accounting department posts timesheets.

13. PMs should inactivate projects, phases, and tasks as soon as they are complete.

In addition to these steps, you will want to document all possible exceptions that can occur, along with the company policy for working less than the required hours in a day, and recording PTO, holidays, and overtime.

While this is an extremely simplified example of an employee timesheet process, you can imagine how haphazard your staff's behavior will get without any documented process. This creates so many slowdowns in other areas of the business and causes the accounting department tremendous grief in chasing people down for timesheets and making corrections.

Even simple changes to your processes can have a profound effect on people's productivity and performance. It is up to firm management to do everything it can to make employees' jobs as easy as possible. With the way that we are all barraged with communications these days, it is easy for up to two to four hours a day of an employee's time to be wasted with useless activities. Investment in analysis and improvement of your business processes can make a tremendous difference when profit margins are low. And there is no greater area that improved processes can have an impact on the firm than in project management as we will discuss later in this chapter.

PROJECT FINANCIAL MANAGEMENT

In Chapter 2, we viewed project management from 30,000 feet and noted some of the areas where PMs get off track in trying to manage the deliverables of their projects, and get them executed profitably. We identified three major categories where project financial management gets derailed in most firms:

- Poor project budgets
- Failure to use resources as planned or adhere to the project scope
- Unexpected changes in the project scope, resources, or outside factors

We have looked at a high level at why money gets lost in our project management processes, and now we need to look at what we can do about it. In order to understand how to fix these problems, we need to understand the challenges faced by our PMs on a daily basis, as well as some fundamentals that help us break down the project management process into manageable components.

Elevating a Project Manager to a Business Manager

As we discussed earlier, in order to improve any aspect of our financial and project management, we need to look at people, processes, and technology. We identified several reasons that PMs are challenged to manage their projects. Now we will take a holistic view of the world of a project manager and what their daily work life involves.

Most PMs are promoted to their position because they have shown some level of technical expertise, had success working with clients, and delivered projects on time and to the client's satisfaction. Very often they may not have had training for their positions, especially in the areas of project management, managing a staff, or dealing with money. Making money on projects tends to be the most challenging area for many managers, and no wonder! The typical PM has a myriad of responsibilities that are thrown at them, often with no training or proven proficiency including:

- Responding to RFPs and creating proposals
- Estimating project fees
- Business development

- Project budgeting and planning
- Project financial management
- Project quality control
- Management of the project timeline
- Billing
- Collections
- Managing subcontractors
- Reviewing and approving time and expenses
- Using resources effectively
- Managing staff schedules
- Maintaining high utilization
- Solving client problems
- Nurturing client relationships
- Mentoring and training staff
- Recruiting and interviewing new hires
- Managing staff performance, and dealing with performance and behavior issues
- Managing contractual requirements and deliverables

I believe anyone can see that stepping into this position without extensive training would be stressful and risky. Any one of these responsibilities would require some level of mentoring, training and supervision to be successful, yet most managers get minimal training besides on the job training.

Training can make a huge difference in a PM's success. In breaking down these responsibilities into what they need to know to be successful, we can start to build a list of key areas to focus in the professional development of a PM. Depending on the responsibilities assigned to your PMs, the following is the minimal level of knowledge they will need in order to be successful at the tasks for which they are responsible:

- Understanding of the basics of how profit is made on projects: labor, expenses, overhead, billing rates, and multipliers
- Relationship of utilization to profitability: how overhead is affected by non-billable time
- Terminology: work-in-process, AR, net revenues, net multiplier, overhead rate, realization, utilization, backlog, variance, etc.
- How to read project reports that are provided to them
- How to budget and plan resources within the scope/fee of the contract
- Operation of the accounting and project management systems used by the firm
- Effectively communicating with the team
- Controlling the subcontractor budget, scope, and billing
- Managing staff, delegating work, human resources (HR) issues, and performance management
- Handling difficult clients and problems that arise during the project (conflict resolution)
- Enforcing compliance with the company's policies and processes
- Keeping up with the latest technical trends and changes
- Proposal preparation processes and best practices

It is no wonder that good PMs are hard to find and even more difficult to keep. PMs who excel in all of these areas are few and far between. They must learn and be able to process a great deal of information every week, and your ability to make their processes more streamlined will provide a great ROI.

If you take a good look at how information is delivered to your PMs, you may see that they are either not getting the data they need, or possibly, they really don't understand it. For example, here is a short list of some of the financial data they need to have access to in order to do a good job in managing their projects:

- Budget to actual by project/phase (real time is best)
- Aged accounts receivable (updated frequently)
- Employee utilization for their direct reports
- Backlog (especially if they are running a team or department)
- Burn rate/earned value (estimate to complete)
- Company overhead rate
- Billing (and preferably) cost rates of their staff
- Project profitability
- Departmental analysis (department heads)
- Forecast of probable wins
- Staff availability

If your firm falls into the trap of not providing detailed financial data to managers, they might be running blind when it comes to really being able to make a difference in the profitability of their projects. Additionally, helping them learn the basics and even the finer aspects of financial management can be even more challenging, given their hectic schedules and overwhelming workloads. Add to that inefficient processes and systems, and you have PMs who are working a lot of overtime and barely able to do what needs to be done to make money for the firm.

Bob Stalilonis, Senior Solutions Architect at Deltek in Woburn, Mass., provides some advice on the use of project collaboration systems to manage the communication issues on a project:

Best practices A&E firms structure their communication to make sure the internal design team, consultants, and the client are on the same page. These organizations use project websites and other communication vehicles (including Kona www.kona.com) to make sure that project information is accessible by all parties and that the information is organized for easy access and comprehension. Establishing a communication plan ensures that everyone is engaged and updated on the project status. Specific tactics include weekly status meetings, monthly progress reports, and issues lists with status/action owner/target resolution date. Many firms provide more comprehensive status reporting and communication requirements on larger projects to mitigate risk as well.

• •

In addition to a communication plan and collaboration tools, here are seven other steps you can implement that will have an impact on improving the effectiveness of your PMs, and their overall productivity and contribution toward profitable project financial management:

1. Implement an employee professional development plan to identify key areas for training and development of business skills

2. Provide training on all areas of responsibility that were identified in the plan including:

   a. Project management

   b. Systems operations and reporting

   c. Financial Concepts

   d. Best practices (best implemented through mentoring)

3. Implement systems with real time access to project financial data

4. Commit to sharing firm financial performance reports

5. Overhaul the firm's operational and administrative processes to be more efficient

6. Provide cost-effective administrative resources to assist PMs with mundane tasks

7. Set clear-cut and measurable goals and performance metrics, and reward excellent performance

By focusing on how you can help make your PMs job easier, and improving their skills around business financial management, you will naturally see a bottom-line improvement to the profitability of your projects. Deborah Gill, CPA, Controller at Clark Nexsen in Norfolk, Va., does a great job of summarizing this concept:

We often assume that our Project Managers are going to assimilate knowledge about project accounting along the way and be able to understand metrics as soon as they don the PM cap. No architectural or engineering school teaches what overhead is, how to calculate a bill rate, or the basic triangle:fee, scope, hours when negotiating a project fee. I recommend you develop a project accounting class for newly minted PMs or at the registered professional level, so they can negotiate a profitable fee from the beginning.

Elements of Project Cost

There are five elements that make up the costs on a project. Understanding and controlling each of these elements is critical to effective project financial management.

1. Labor

Labor is the largest expense of both an A&E business and its projects. In order to manage labor costs, the A&E firm must put controls in place to ensure that each employee is:

- Working on what they are supposed to be working on
- Filling out their timesheet accurately
- Maintaining target utilization rates
- Adhering to the scope of each project they are working on

The labor costs charged to the project will include the employee's raw labor cost, which is calculated as their salary divided by 2,080 hours per year, multiplied by the number of hours charged to the project on their timesheet. A relative amount of indirect costs such as fringe benefits and overhead will follow these costs and be applied to the project to calculate true project profitability (see more on this later in this chapter). In the case where salaried employees work more than the number of hours in the pay period, their effective hourly cost must be diluted and spread among all the projects on which they worked in a pay period to calculate true project cost. When employees work a lot of uncompensated overtime, this has the effect of lowering the cost of the labor to the company. In federal government contracting, the process of diluting costs and applying the effective cost to the project is called *total time accounting* and is required for accurate job costing and billing.

2. Expenses/Other Direct Costs (ODC)

Based on the terms of the contract, expenses charged to a project may be reimbursable or direct. Reimbursable expenses such as printing and travel are able to be billed in excess of a fixed fee, or are reimbursable as part of a T&M contract. Direct expenses are included in the total fixed fee on the contract.

Costs must be captured internally and charged to projects accurately. If this process is not well-organized, reimbursable expenses can slip through the cracks and end up not getting reimbursed. Direct expenses should always get charged to a contract, and accounted for when estimating a fee for a project. If actual direct expenses exceed the amount estimated on the fee project, it is usually difficult to get them reimbursed. It is essential to control these costs and more importantly, estimate them correctly in the beginning.

Most firms have charges coming from vendors, internal charges such as blueprints that are often tracked electronically or on a log, and employee expenses. Processes should be developed to ensure that expenses are captured and charged to projects accurately, and that employee expenses are approved.

3. Consultants/Subcontractors

For many A&E firms, subcontractor costs can be up to 50% or even more of the contract value. Managing subcontractors requires carrying out many administrative and project management tasks, including negotiating contracts, managing the subcontractor work, and paying subcontractors correctly.

Most subcontractor costs are negotiated and controlled by firm principals or PMs. In some cases, there may be a markup included in the fees billed to the client, and in other cases, a subcontractor management fee is added to the actual cost of the subcontractor. In most cases, subcontractor fees are pay-when-paid (PWP), meaning that the subcontractor does not get paid until the client has paid the prime contractor, or the company holding the main contract. Managing PWP payment terms can be cumbersome and time-consuming, and requires careful management of billing, cash receipts, and accounts payable processes.

From the PMs standpoint, subcontractor invoices must be checked against negotiated contract fee amounts, as well as determining if the subcontractor is billing the correct amount based on the work performed or percentage completed of the negotiated work. You may find that a purchase order (PO) system can help control your subcontractor costs, especially if you have problems with subcontractors billing you the wrong amounts, sending duplicate invoices, or billing more than the contracted amount.

4. Indirect Expenses (Fringe/Overhead /G&A)

In order to accurately calculate project profitability, you must apply the indirect costs of running the business to your billable projects. Indirect expenses are captured in overhead projects and the financial accounts of the business. Indirect expenses include fringe benefits (medical insurance, PTO, etc.), non-billable labor costs, rent, company insurance, supplies, marketing, interest expense, and all other expenses that are incurred to keep the company running and are not directly attributed to a project.

For the purpose of this discussion, we will call all of these indirect expenses *overhead*. Many A&E firms lump all of their indirect expenses into one bucket, so this is a fairly conventional way to analyze indirect costs. However, for firms doing federal government contracts, the indirect expenses may have to be captured and accounted for in separate "buckets" called *cost pools*. We will look at accounting for government contracts in more detail in Chapter 9.

Overhead rates are usually calculated by dividing the total overhead costs by the total direct labor costs for a period, and are reported cumulatively on a year-to-date basis. Some companies estimate their overhead costs at the beginning of the year and compare them frequently. This is a recommended best practice, and as we will see later in this chapter, can help avoid severe losses on all contract types. Losses will occur when actual overhead rates exceed the estimated, or provisional overhead rate used to form the billing rates and estimates used to bid on projects.

Achieving profitable business operations requires managing and controlling overhead costs. Some of the primary reasons that overhead is higher than estimated include:

- Increases in non-billable/indirect time (poor utilization)
- Write-offs
- Employee growth and turnover
- Not controlling expenses
- No company budgets
- No ROI on assets
- Pursuing projects you can't win
- Slow-paying clients
- Unexpected purchases or other expenses

5. Contingencies

Contingencies are unexpected costs that are very difficult to estimate and plan for. As we discussed in Chapter 2, a contingency amount should be included in the project estimate and budget based on prior history of the firm, and for the specific client or type of project. Common contingencies include:

- Client changes
- Unexpected problems with site, regulatory, or environmental issues
- Creating a scope with an unattainable approach
- Employee or client staff turnover, conflicts, and changes
- Delayed project start or funding/financing issues
- Slow-paying clients
- Subcontractor issues

Evaluating past project data can provide some insight into how to estimate contingencies on a given project or with a particular client. The more you have worked with a team, the easier it should be to figure out how to plan for unexpected changes and circumstances.

Contracts

Nailing down the scope with the client before the contract is signed is critical to project success and profitability. For clients with whom you have a long history working, this might be easier. Proposals and even contracts may be as simple as a letter, outlining at a high level the scope and agreed-upon fees.

But in a more competitive process, working with clients with whom you do not have a strong relationship or extensive working experience, it is essential to spend a great deal of time communicating detailed expectations and ensuring that these expectations are fully documented. Some best practices that can greatly add to the success of a project include:

- Implementing a defined process for approval and billing of extra services and changes to the scope

- Contractually agreeing upon the number of meetings and reimbursement for extra meetings required by the client

- Defining how contingencies will be handled, including unexpected project expenses, increases in fuel and materials costs, the effects from unexpected government regulation or interference, etc.

- Including contractual clauses that guarantee reimbursement for costs due to substantial project delays.

- Defining billing terms and finance charges for late payment of invoices

- Inserting a limited liability clause reducing the firm's liability to net fees in order to keep insurance costs down

- Requiring payment of retainers before projects begin, along with instructions on how they will be applied to invoices.

- Building in rate increases for multi-year contracts to ensure you can increase labor wages each year without losing money.

- Breaking up the project into as many small pieces as possible, which allows you to easily make changes, and add or subtract "options" as needed.

Contract Types and Project Management

The type of contract can also have a big impact on the profitability of and the difficulty in managing a project. The two main types of contracts are fixed amounts and cost based.

1. Fixed Amount Contracts (Lump Sum)

Fixed amount contracts take different forms, including firm fixed price, fixed fee, or phased fixed fee. They are also commonly called *lump sum contracts*. For the most part, they are all the same except for the phased fixed fee contract where the fixed amounts are broken down to the phase level. For the purposes of this discussion, I will refer to them as fixed fee contracts.

When estimating a project that is to be billed on a fixed fee basis, the cost should be calculated the same way it is done for a cost based contract. First, labor is calculated by taking the hours estimated by phase and multiplying by the budgeted rates for the labor categories required, and then calculating the total cost. Estimated expenses, subcontractors, indirect expenses, contingencies, and desired profit are added to the labor estimate to determine the fee amount that the firm will bid.

Fixed price contracts have an inherent level of risk built into them. If you spend more than what you bid, you will not reach your estimated profit margin. You can even lose money. It is critical that PMs have access to real-time project expenditure data to closely control what is happening on their projects. Fixed fee projects are the type most likely to lose money, and PMs need to be given the tools to stay on top of them easily, before budget overruns occur. Best practices for managing fixed fee contracts are as follows:

- The scope must be very narrowly defined in the contract, and it must specifically describe what is and isn't included in the fee.
- The scope must be closely monitored, and any scope changes must be negotiated and approved in writing. This requires excellent communication among the project team.
- Short-term projects must be managed almost daily as they can overrun quickly.
- Timesheet approval is critical in order to ensure that employees are doing what they are supposed to be doing and not performing work outside the scope or additional services.

2. Cost Based Contracts

Cost based contract types include T&M, also called hourly, and cost plus fixed fee (CPFF). T&M or hourly contracts are billed at set billing rates for specific labor categories (principal, senior engineer, CAD operator, etc.). Cost plus contracts are billed at the raw labor cost of each employee, plus a markup or burden for fringe benefits, overhead expenses, G&A expenses, and a profit fee. Cost plus contracts are usually only required for government contracts, and may also leave your firm open to potential audit. See Chapter 10 for a more detailed explanation of the issues in managing cost plus contracts.

In some cases, T&M contracts will have a not-to-exceed limit placed on them. This provides a serious advantage to the client, in that if the charges are less than estimated, they pay the lower amount. If the actual charges exceed the limit, then they don't have to pay more than the upset limit. These types of projects must be monitored closely and managed for scope creep just as carefully as if they were fixed price.

While T&M contracts generally have a lower risk, many PMs believe that the company can't lose money on T&M contracts, and sometimes not enough scrutiny is given to managing the charges, write-offs and scope. The billing rate used on the estimate and contract is based on estimated costs for labor, fringe benefits, and overhead. Therefore, if the actual overhead rates incurred by the company exceed those used in the estimates, the firm could be losing money on every hour billed.

In looking at how the labor billing rate is formulated, we can better analyze how to control project costs. A bill rate is calculated by taking the direct costs, including the labor and any DPE such as fringe benefits, plus overhead, G&A, and a desired profit margin.

Figure 4 illustrates how a typical billing rate is calculated. You can see from the example, that if the actual fringe benefits or overhead rates are higher than the rates used to prepare the estimates and budget, then the company will not be recovering the net multiplier it expects. For this reason, it is essential that indirect rates are reviewed frequently, and templates used to create project estimates are updated accordingly.

How Billing Rates Are Calculated			
Estimated		**Actual**	
Direct labor (DL)	$40.00/hour	Direct labor (DL)	$40.00/hour
Fringe/DPE (20%)	8.00	Fringe/DPE (22.5%)	9.00
Overhead (100% of DL)	40.00	Overhead (110% of DL)	44.00
Total costs	$88.00	Total costs	$93.00
Target profit margin 15%	13.20	Actual profit margin 7.5%	$7.00
Target billing rate $101.20 (usually rounded to $100)			

Figure 4

In this example, the net multiplier, which is the bill rate divided by the direct labor rate, is as follows:

$$\$100 \div \$40 = 2.5$$
(this means that the firm is billing 2.5 times the employee's rate)

As you can see in Figure 4, the issue with T&M contracts is that in order to make the target profit, the actual indirect costs must be in line with the estimated indirect rates used in the estimate (fringe and overhead). If the actual fringe and overhead rates are higher than the estimated rates by the amount of the target profit percentage, the company does not make the estimated profit, or it loses money. Indirect rates used in estimates should be updated frequently to ensure that PMs are using updated rates, and billing rates should be analyzed at least annually to ensure that they are still appropriate.

With the profit margin narrowed down to 7% of the billing rate, there is more room for other problems to have a negative impact on profitability. Strict policies should be put in place for writing off or holding time each month. PMs should have to justify holding or writing off hours, and they should be approved and scrutinized carefully. Monitoring potential billable charges to actual billed charges, as well as monitoring an aged unbilled work-in-process (WIP) report, can provide an indication whether PMs are doing a good job managing T&M contracts.

PROJECT ESTIMATING

As explained previously, correctly estimating a project, before you have won it, is a critical step toward ensuring the success of the project. It creates the basis for the scope of services agreed upon in the contract, as well as the budget your firm will use to manage the project once it is awarded.

A good estimate has the following components:

- Breaks down the work into phases and tasks in the order that they will be started
- Determines the skills levels and labor categories needed to accomplish each task, as well as where subcontractors will be required for work the firm can't do
- Calculates the estimated hours that it will take to accomplish each phase and task
- Applies budgeted hourly rates to each hour by labor category to calculate the cost of labor
- Estimates the expenses required, including travel, blueprints, delivery, etc.
- Estimates the cost of subcontractors on each phase and task
- Uses an implied overhead rate and target profit percentage to burden the direct costs
- Includes a line item or multiple line items for contingencies
- Includes a line item for meetings for each phase
- Calculates the total estimated fee once all the costs are estimated

You will notice that we have not assigned specific people to the proposed project at this stage. We often do not know what resources we will actually assign to the project at this point. We probably won't be able to assign dates and even timelines to the project at this point. We may try to estimate some start and end dates. But so many things can throw off the start of a project, so we are usually forced to do a ballpark guess at this point.

While this illustrates the critical components at a high level for preparing project estimates, I find that many firms do not have an established process for ensuring consistency across all estimators. This is also a common issue with proposals. This creates a number of issues that later flow down through the contract, project budget, time management process, and project reporting process.

The following are several areas where inconsistencies are often found in the estimating process that can cause issues for the project manager after the project is won:

- Using old cost and billing rates
- Using spreadsheets with inaccurate formulas
- Using last year's overhead rates
- Estimating cheaper people than is required to do the work effectively
- Failure to consider contingencies
- Failure to use past history on projects to accurately estimate hours
- Using different language between estimates to describe phases and tasks
- Underestimating subcontractor costs
- Failure to estimate the amount of meetings that will be required (and the time required to prepare for them)
- Using different formats/spreadsheets to create the estimates
- Not managing the estimates in a central repository for global access

- Not tying the estimates back to the project once it is awarded
- Failure to get estimates approved by upper management before submission

The answer to this problem is establishing a systematic approach to preparing cost estimates. This can be done with spreadsheets, as long as there is tight control over them. Spreadsheets can be risky and fraught with errors. If you are using spreadsheets, you need to ensure consistency, accuracy of data in the templates, and access to them by upper management. By using well-defined templates that are distributed by accounting staff, you have a better chance of eliminating many of the issues noted above.

I am not a fan of using Microsoft Excel for estimating and budgeting. In my blog article, *Three Reasons Companies Should Outlaw Excel* (Source: http://acuitybusiness.com/blog/bid/175066/Three-Reasons-Why-Companies-Should-Outlaw-Excel), I argue that Excel spreadsheets are fraught with errors, waste a lot of time, and are unique to the creator to the extent that no one else can effectively manage them.

A more effective method of estimating is to use an integrated system that has a structured process for creating project estimates and integrates with your accounting system. By using an integrated system, you will be able to develop a much more streamlined process, and ensure better accuracy and greater visibility by upper management. While project planning software can be somewhat difficult to implement, the benefits include creating a lot more structure and process around the way your managers develop estimates. Ultimately this will result in more accurate estimates and better results

PROJECT BUDGETING

The estimate that is developed during the bidding process is the foundation for the contract, and resulting budget. The project budget is the company's realistic expectation of how the project will proceed and what it will cost to get it done. It may or may not coincide with the amount agreed upon in the contract. The budget should combine the scope, estimate of costs as determined in the bidding process, as well as any other pertinent information such as size, complexity, estimated timeline, expected budget changes, and adjustments to account for client expectations.

Inevitably, your budgets will change over time, as the project scope is adjusted to either add or subtract components to the amounts agreed to in the original contract. Contract modifications will need to be agreed to and tracked internally to ensure that you are billing your clients accurately.

An effective budget has the following components:

- A work breakdown structure that includes all required phases and tasks in the order to be completed
- A realistic start and end date for each phase and task
- The people or labor categories to be assigned to each phase and task with associated billing rates
- The estimated hours spread along the timeline. The closer the hours can be assigned to the time period in which they are expected to be spent, the better.
- The expenses expected to be incurred, also by pay period or month
- Subcontractor costs expected to be incurred by phase, task, and month
- Line items for additional services added to the scope after the project is started
- A line item for contingencies
- An estimate for meetings within each phase

As the budget is created and agreed upon internally, it is critical that you monitor and track the actual costs you are incurring against the budget line items. The key to good project budgeting practices is not only in the original development of the budget, but also in what happens after the budget has been established and the project is being executed. Some firms have excellent processes for tracking progress along the project schedule, and providing detailed and timely information to PMs as the project progresses. Other firms drop the ball in this area, failing to give managers real-time information and alerts as spending milestones are hit.

If possible, project budgets should be monitored on a weekly basis. The more often a PM can see his or her actual expenses against the budget, the better chance that PM will have to make adjustments and get the project back on track. An automated project budgeting system can help a PM outline his initial budget and keep track of changes. The ideal system will integrate with the timesheets in order to see how the project is tracking as these adjustments are made. Assigning resources to each level of the work breakdown structure on the project can help the manager control who is charging to this project and whether they are working on the assignments they have actually been assigned.

There are two ways to approach the budgeting process: top-down, and bottom-up. A top-down budget is one that is created by allocating the total expected fees to individual phases on a percentage basis. Many firms utilize this process of budgeting in order to come up with a baseline projection of fees. However, it is recommended that a bottom-up budget be created and compared with the top-down budget to obtain a reality check. A bottom-up budget is where specific resources are assigned to the lowest level of the work breakdown structure at budgeted rates. By assigning specific employees or generic resources to the tasks on a project, PMs can better monitor project progress.

After the resources have been assigned to the tasks, PMs can start to assign hours to specific time periods. As the costs of each level roll up, the total value of the project budget is calculated. By assigning hours to specific time periods, the firm will be able to forecast revenue, as well as track whether employees are performing as planned or spending money too fast.

For best results, the manager should be monitoring the budget to see the following:

- Who is charging to the project, and were they scheduled to work on it?
- Are they charging to the correct phases and tasks?
- Are they charging more hours than were planned for the period?
- Is the money being "burned" faster than was planned?
- Are the employees and subcontractors sticking to the contract scope?

It is critical that accurate rates be used in the budgeting process. When the project budget is integrated with weekly time management, optimal project management reporting is available. Without regular real-time access to this information, you are more likely to have budget overruns.

In addition to budgeting labor, subcontractors and other direct project expenses must be budgeted. Optimally, you will want to budget subcontractors at the firm or vendor level so that you can track how your subcontractors are invoicing you against their contracts.

RESOURCE MANAGEMENT

Managing your resources in order to maximize the value of each employee's efforts is critical to profitable business management. There are three basic components to ensuring that your resources are being utilized as effectively as possible.

- Careful planning, monitoring, and adjusting of employee assignments into the future
- Weekly adjustment of assignments based on project progress, delays, and changes to the scope
- Effective time management, including a diligent approval processes

The estimating and budgeting process will start building the framework of your resource scheduling. By planning when you expect your resources to start and finish specific tasks, you can now start to monitor actual time charged against the plan.

If some of your employees are overscheduled and others don't have enough work to do, employee time is not being optimized. Reviewing the entire company's schedules will allow you to more easily see where your employees are being better utilized.

One of the problems with effective resource management is that certain skilled employees may be in higher demand than others. In most firms, employees work on many different projects, and may be shifted from project to project during the week. Managing and keeping track of all of these schedules is very difficult. In addition, specific skills may be required, and employees needed to complete specific tasks may not be available. This is a very challenging process for most PMs. As we discussed earlier, balancing the need for high utilization with project profitability can be difficult.

Implementing and automating a resource management scheduling process can allow managers to more easily make adjustments as resource conflicts arise. Optimally, your managers should have a way to search for resources by labor category or specific skill sets, and they should also be able to see which resources are over- and underscheduled. By providing your managers with a way to see your resources by department, office, and project, you will better equip them to maximize use of available resources.

In addition to having your resources optimally assigned, it is also beneficial to be able to forecast resource requirements into the future. This can help you make better decisions in hiring or reducing your workforce. Knowing about specific backlog shortfalls into the future can help your company make better management decisions.

By putting effective tools in the hands of your PMs, you will enable them to maximize the profitability of each project and the firm as a whole. While it can be difficult to schedule every project in your company, there are ways of doing it that can simplify the process. A resource management tool can improve the quality of your scheduling process, as well as the efficiency of your Monday morning meetings, by allowing a visual method for evaluating staff assignments and projected requirements.

Another cost-effective practice is utilizing administrative staff to assist in keeping budgets updated and making scheduling changes. This can help reduce the costs of your PM's administrative time in tracking schedules, as well as ensure that data are up to date and consistent among managers, offices, and projects. You will also find that it saves money in training PMs, and increases their productivity and job satisfaction by reducing their administrative responsibilities.

CONTROL OF EXTRA SERVICES

Extra services are where a lot of A&E firms lose money on projects. Effective project management requires careful adherence to the project scope and budget, as well as attention to what employees are doing and where they are charging their time. However, because of the culture traps that are so prevalent in many firms, many employees choose to ignore the scope in order to ensure quality and keep the client happy. Controlling extra services requires careful planning and processes, including the following recommended practices:

- Agreement with clients, including detailed descriptions of how extra services are to be handled in all project contracts.
- Develop a company policy and process for management of extra services, including development of contract modification documents, approvals, and system management.
- Train employees on the process for handling extra services requests from clients. This can be intimidating to some staff, and having a good process can make a big difference.
- Require that all extra services requests get approved by the project manager and client before the employee does the work.

- Create special codes and phases or tasks in your project work breakdown structure to separate extra services and track them separately.

- Communicate the entire scope to the project team at the beginning of each project.

- Closely monitor timesheets to ensure compliance with the extra services policies.

- Review and enforce compliance with policies on a regular basis.

Ed Friedrichs, former CEO of Gensler, has an excellent suggestion for handling services requests from clients who are outside the scope of the contract:

I found that the nature of the conversation with the client about additional services shaped the outcome. It generally went something like this—"What you're asking us to do (note that I place the request on the client's shoulders and present it before we've done the work) is outside our contract scope. This is why it has come about…, this is the cost of doing it…, these are the implications of not doing it… I wanted to give you the option of either not going forward, or authorizing our additional scope." The discussion usually went well using this approach.

By carefully planning and training staff on how to manage requests that are outside of the scope of the contract, a great deal of project budget overruns can be avoided, and project profit margins can be significantly increased. As time management is critical to ensure adherence to the contract scope, improving your timesheet processes will have a major impact in this effort. The next chapter investigates how you can vastly improve your timesheet processes to impact project success.

Summary

- Standardized processes and procedures are necessary to ensure that your employee's time is utilized efficiently, and projects are managed consistently. Processes should be documented, and employees should be trained on how to follow them. Strict enforcement of policies will improve operations, ensure quality work product, and maximize project profitability. A detailed explanation of how to improve processes was provided.

- Careful attention to your firm's project financial management practices can result in increased profits and happier clients. PMs have many responsibilities and are not often trained adequately on many areas of their jobs. Training managers on financial management concepts, employee management, communication skills, and systems can cause a significant improvement in their performance.

- There are two main types of contracts: cost based and fixed fee. In calculating profitability for each type, you must analyze the labor, direct expenses, and consultant costs, as well as overhead. If actual overhead is higher than the rates used in your estimates, you can lose money on the contract. In addition, write-offs of time on a T&M contract or inability to bill for your employee's work can also erode project profits.

- Project estimates are the foundation for a project because they impact the scope, contract, and budget. If you have inconsistent estimating practices or utilize poor estimating templates, your projects may be destined to lose money from the start.

- The project budget should be developed at a detailed level outlining the phases and tasks to be performed, as well as the resources and hours estimated to get the work done, spread out over time. Managers should be monitoring actual costs each week against the budget line items to ensure that problems and overruns are caught early, and adjustments are made quickly to avert additional losses.

- Managing project resources to achieve target utilization rates is critical to the profitability of an A&E firm. This involves weekly managing of employee schedules to ensure the optimal workload for each employee, as well as forecasting labor requirements into the future to enable better hiring decisions.

- Many firms lose money because they are not able to bill for extra services. In most cases, this is because of scope creep which is a failure by your project team to stick to the original project scope, and perform services outside of the scope without getting agreement from the client to bill them as extra services. This is often a result of the culture traps discussed in Chapter 1. In order to ensure that extra services are recovered, adequate processes must be put in place to educate the project team about the scope and ensure that all extra services requests are approved by the client before the work is done.

CHAPTER - 6

TIME IS MONEY

"Time is more valuable than money. You can get more money, but you cannot get more time."

~JIM ROHN, AMERICAN SPEAKER AND AUTHOR

In a professional services business, the expression "time is money" is truly accurate. Your employees' time is the greatest asset your business has. It needs to be managed with great care, as if each hour of their time is a valuable piece of inventory. Certainly, if you were selling diamonds for a living, you would build a secure storage facility, get armed guards, and buy insurance to protect this valuable asset.

However, most firms do not protect their greatest asset—employee time—with this level of care. Employee time is often the most ignored asset in the company. Employees are notorious for complaining about having to do timesheets and often look at it as a cumbersome requirement. I regularly hear that firm principals, even ones who are highly billable, are the biggest violators of the firm policies, and cause the accounting department extra work and little support in trying to get employees to submit timesheets on time.

The reason that timesheets are so critical to your business performance is that they pervade almost every single facet of the business. Timesheets affect all of these areas:

- Project profitability and reporting
- Billing
- Payroll
- Financial statements
- Overhead rate calculation and allocation
- HR (benefit accruals)
- Utilization reporting
- Revenue recognition
- Project management effectiveness
- Scheduling
- Employee performance analysis and compensation
- Project Scope control and ability to bill for work outside the scope (extra services)
- Government contract compliance and reporting
- Client satisfaction and trust in your billing and reporting capabilities

As you can see from this list, timesheets affect many critical aspects of the business. Tightening up loose processes and enforcement of better policies will have a profound effect on the entire firm including:

- Reduction of corrections during the billing process, which will shorten the billing and cash receipt cycle
- Reduction of budget overruns
- Increased project profit margins due to better resource management and billing of extra services

- Better business decisions due to improved project reporting and analysis
- Improved bidding and estimating due to more accurate historical data
- Improved employee performance reporting
- Better HR management
- Compliance with government contract requirements

How can you do a better job to enforce timesheet policies and ensure that timesheet entries are more accurate? Below are some best practices that I have learned over the last 24 years that have been successful for many of our clients. In evaluating your current processes, you can develop better methods for implementing timesheet policies, enforcing these policies, and controlling and improving employee behavior.

- **Develop a documented timesheet entry, editing, and submission policy.** Employees should be given a detailed document that explains the company's timesheet processes. For new employees, this should be reviewed and emphasized during the onboard process. It is best if these documents include screenshots of the firm's timesheet system with a description of how to fill in each field. This document should include a detailed description of all aspects of the time entry process, including making changes to timesheets and the company's submittal and approval processes.

- **Teach employees how critical timesheets are to the entire profitability of the firm.** It is amazing how many employees do not make the connection between their timesheets and their paychecks. By spending the time to educate your staff about how the firm makes money, and how crucial accurate and timely timesheets are to the firm's profitability, they will become more sensitive to requests for compliance to the firm's timesheet policies.

- **Provide user training on timesheet entry for everyone.**
  Training on use of the firm's timesheet system should be provided when the employee is hired or when processes are changed. I recommend recording the training as a webinar that you can have employees watch. Then provide special discussion time afterward for questions.

- **Train supervisors and PMs on timesheet approval processes.**
  Timesheet approvals are critical to good time management and may also be mandated on government contracts. Timesheets should be approved by supervisors for verification of hours charged, regular versus overtime hours, charges for PTO or holidays, and overall compliance with the company timesheet policies.

  PMs should approve the timesheet to ensure that the employee is charging to the correct project, phase, and task; that they have not exceeded the numbers of hours planned; and that they are not outside the scope and working on extra services that require a contract modification and could be billed separately from the base contract.

- **Implement line item time approval.** Timesheet line item approval allows the PMs to approve time by project rather than by employee timesheet. This is a much easier way for them to see all charges by all employees to a single project, and reject individual time charges for correction or transfer to another project or phase.

- **Configure the timesheet to automatically come up in the morning when the employee turns on their computer.**
  This may be possible in some systems. Check with your information technology (IT) professional to determine if they can make this work.

- **Utilize a mobile application for timesheet entry.**
  Compliance with timesheet policies will be improved if you make it as easy and convenient as possible. Many systems now offer a mobile timesheet application that allows employees to enter time from any location or device over the Web.

- **Require daily time entry and enforce timesheet policies.**
  This is my number-one recommendation for effective time
  management. Regular daily time entry can make a major
  difference by improving timesheet accuracy, billing accuracy,
  and cash flow. If a professional services firm can do only one
  thing to improve business profitability, I would urge them to
  require electronic daily time entry by all employees.

  I share this with our clients regularly, and I often get the
  response that this is an impossible process to implement. The
  firm's culture and unwillingness to enforce company procedures
  is one of the major reasons that our clients continue to suffer
  with inaccurate timesheets, lost extra services, as well as slow
  and incorrect client billings. This is actually the easiest and
  most cost-effective change you can make immediately which I
  promise will produce instant positive and measurable results.

- **Submit and approve timesheets weekly.** If you have a weekly
  or bi-weekly payroll process this is easy to implement. It is
  harder with a semi-monthly payroll process but can be done
  depending on your billing cycle. Weekly time approval will
  force PMs to look at what employees are doing more often,
  and alert them to problems and scope creep on a timelier
  basis.

- **Limit the number of choices employees have on their
  timesheet.** It is human nature that if you give someone a lot
  of choices, they will get overwhelmed or confused, and end up
  guessing. Limit the number of items to choose from in each
  field on the timesheet. By reducing the number of choices,
  their final selections are likely to be more accurate.

- **Restrict employees from charging to projects that they
  are not allowed to charge to.** If your system allows, block
  employees from charging to projects on which they are not
  authorized to work. This will reduce the number of choices
  they have and increase the likelihood that they will pick

the right project. This can also be very effective if you have employees in different offices or departments and don't want them charging to the other team's projects without official approval. If you utilize reimbursable phases or tasks on your projects, this can also keep employees from charging their time to codes that are obviously not meant for labor charges.

- **Inactivate phases/tasks under the project when they are completed.** By inactivating phases and tasks that have been billed 100% complete, you can keep people from continuing to charge to them. If possible, limit them to having access to one or two phases at a time. This will help increase accuracy and potentially capture extra services by alerting the employee to out-of-scope issues.

- **Require PMs to budget projects at the phase/task and employee levels.** In order for a PM to determine if an employee is on track with a project plan, he or she needs to budget at the lowest level of the project to which the employee will charge his or her time. A good project planning system will have the capability to send instant timesheet alerts if an employee is exceeding the budgeted hours, or charging to a phase or task to which he or she is not assigned.

- **Integrate project resource assignments with timesheets.** If your system allows, provide project plan data within the timesheet to show employees where they are assigned while they are filling out their timesheet. Some systems will even allow you to have the project plan populate the project/phase/task data on the timesheet, which will further improve accuracy.

- **Implement workflows and alerts.** If possible, have your system warn PMs about project overruns, employees charging to the wrong projects phases and tasks, or charging too many hours to a specific activity.

- **Alert employees when timesheets are due to be submitted.** If possible, provide advance e-mail alerts to remind employees when timesheets are due to be submitted, and send alerts when employees do not fill out or submit their timesheets on time.

- **Implement an incentive for submitting timesheets on time OR make employee's bonuses and/or other incentives dependent on strict adherence to company policies.** While this is not a commonly used process, I have seen this work very effectively to incentivize employees to follow timesheet policies. One firm I worked with saw their timesheet violations reduced to almost zero by promising to reduce an employee's bonus by $100 for each day they do not fill out their timesheet on time. While this can be difficult to monitor and may be seen as an overly oppressive consequence for timesheet non-compliance, it may be necessary in firms where strict compliance is required by law, such as with government contracts, or where timesheets policies are frequently violated and nothing else seems to work.

- **Designate a person to check timesheets daily.** If your system allows, have someone check to see if everyone is doing their timesheets every day. I fondly call this person the timesheet Nazi, and while this is not an enviable job to have, it does have an effect on obtaining the desired results.

One of my long-time clients recounted this humorous story, which illustrates the value of having someone visibly walking the halls of the office (the names have been changed intentionally):

* *

I have a story about "Jingle Frank." Frank was the president of our company, and he would jingle the change in his pocket as he wandered through the office. This made everyone hunker down and work harder when they heard him. He retired 10 years ago, but the stories about him persist. I still talk with him occasionally. I need to ask him if he intentionally jingled the change in his pocket. I suspect his answer will either be "Yes" or "You'll have to draw your own conclusions" (with that grin on his face and in his voice!).

• •

I also believe it is critical that the firm's leaders are frequently communicating to the staff about the importance of complying with timesheet (and other company) policies. If the employees notice that the firm executives are not concerned with timesheets, they won't be either.

- **Implement an automated floor check process.** Automated floor check systems are available in some timesheet management systems and can be programmed to work in others. Effectively, they check if employees have done their time each day and send a notification if they are delinquent. In some cases they can provide a report that flags employees who have violated the timesheet policies more than three times. These are very beneficial for government contractors who are required to record time daily and can be surprised by auditors at any time.

- **Remove an employee from payroll direct deposit for non-compliance.** By threatening (and following through) with removing an employee from payroll direct deposit if they fail to submit timesheets on time, you can have a big impact on employee timesheet compliance. While it can also be a hassle for the accounting department, it is a very effective deterrent, and it rarely has to be done more than once.

- **Take the employee's paycheck to Wyoming when they don't comply.** OK, so this seems really far out, but I got this idea from my long-time client Ed Kroman, who actually did this many years ago. When one of the company's most non-compliant employees failed to complete his timesheet too many times, Kroman took his paycheck on vacation with him to Wyoming and took a picture. Now all he has to do is show the photo below to new employees and he rarely has a problem with compliance from new employees.

Figure 5:
(Courtesy of Ed Kroman, Director of Finance, Summer Consultants, Inc., McLean, Va.)

It might feel like taking an employee's paycheck to Wyoming is a bit drastic, but it really makes an impact on your staff about how important their timesheets are to the company. Getting this point across to them is an essential first step to obtaining consistent compliance to your policies.

For a labor-based organization, every minute counts. Once time is gone, you cannot get it back. How many of you can remember what you did yesterday in 10- or 15-minute increments? (I can't!) Time is inaccurately recorded or wasted with excessive meetings and poor management of the contract scope, or by allowing employees to record time incorrectly because they are filling timesheets out weekly, bi-weekly, or even worse, monthly.

This quote from the Law Biz Blog in December, 2008 shows the potential losses from poor time control:

• •

One missed 10th of an hour each day translates to **23 lost hours a year**. *And failure to keep current, proper time records will usually result in more than just 1/10th of an hour lost ... DAILY ... and MANY thousands of dollars in lost billable revenue! (Source:* http://www.lawbizblog.com/2008/12/articles/ cash-flow-finances/time-sheet-records-revenue-billed/*)*

• •

In addition to the losses from unrecorded or inaccurate recording of time, other critical business management issues are at stake. Timesheet accuracy affects billing accuracy and cash management. It can also have a profound impact on the viability of the company as a whole if you are a government contractor and subject to a Defense Contract Audit Agency (DCAA) or Federal Acquisition Regulations (FAR) audit.

By developing, implementing, and enforcing effective timesheet management processes, you are taking a major first step toward finding the lost dollars that are slipping away every day right in front of your eyes.

Summary

- "Time is money" is an especially relevant metaphor for professional services firms. Your employee's time is your most valuable asset, and careful guarding of it will ensure more accurate reporting and more profitable projects.

- It is critical to develop detailed documented processes for your employees to record and adjust their time in order to ensure accuracy, and maximize billable hours.

- Many best practices are outlined in Chapter 6 that should be included in your company's time management documentation. Daily entry of time is the best way to ensure accuracy, improve cash flow, and keep the company compliant with government contract regulations.

- Approval processes can ensure that time is charged to the correct projects, phases, and tasks, improve billing accuracy, job cost reporting, and employee performance analysis. It can also help managers stay focused on who is working on their projects and avoid having too much time accumulate on phases without their knowledge or control.

- Once time is gone you can never get it back. Time management is one of the fastest and most effective ways you can improve your bottom line.

COMPANY FINANCIAL AND CASH MANAGEMENT

---

*"Clients that pay slow are usually the ones
we lose the most money on."*

~EDWARD (ED) J. KROMAN, III, DIRECTOR OF FINANCE,
SUMMER CONSULTANTS, INC., MCLEAN, VA.

Business financial management does not come naturally to most A&E professionals. This is not a subject typically taught in architecture or engineering schools, and most A&E business owners have to learn it by watching others, or hiring outside CPAs or other financial consultants. Those who came before you may not have been trained themselves, and so the problem of bad habits, lack of discipline, and ignorance as to the importance of critical financial management practices is continued through generations of business owners.

If you are the exception to this legacy, then I commend you. Excellent resources are available to A&E entrepreneurs, specifically geared toward financial management best practices. Trade associations such as the American Council of Engineering Companies (ACEC), and the American Institute of Architects (AIA), have been a great help to A&E principals over the years. Other organizations such as ZweigWhite, PSMJ, PMI, Rust O'Brien Gido + Partners (ROG), and many individual experts are available to provide guidance to business owners on subjects including ownership transition, project management, and mergers and acquisitions.

Your outside CPA can also be a great resource for helping ensure your firm is in compliance with the myriad of state, local, and federal government regulations, and tax compliance requirements. If you do business in more than one state, keeping up with the different payroll, HR, and income tax regulations alone can be overwhelming.

If you are planning to grow, it is essential that you have a trained accounting resource that is helping you manage the finances of the business. I can't even count the number of times that I have seen major mistakes made that have cost my clients a lot more money to fix than they would have spent to prevent them in the first place.

A good accountant is worth his or her weight in gold by ensuring compliance with government regulations, providing accurate financial statements to the owners, dealing with banks, managing cash flow and collections, purchasing insurance benefits and policies, and establishing good internal controls to avoid fraud and embezzlement. Payroll rules alone are complex, and often change every year. If correct rules are not followed, the government can levy significant tax penalties and interest that can be enough to shut the company down.

I share this philosophy with one of my long-time and well-respected clients, Anthony (Tony) J. Vitullo, CPA, CGMA, and CFO of Cooper Robertson Partners, New York, N.Y.:

• •

Having the right financial manager is like having another partner in the business, and this one is constantly looking out for your best interests and that of the firm. Think you are too small or can't afford a CFO? You can't afford not to have the answers to the complex issues presented above, nor can you afford to not have someone who can provide you with the data to run your business and make informed decisions. There are varied solutions to making sure you have the right information and the right financial manager at your side to help you run and grow your business.

Tony has a lot of great insight on this subject, which you can find in his blog article at http://acuitybusiness.com/blog/bid/222581/The-CFO-as-a-Critical-Partner-to-your-Business.

CASH IS KING

Without careful monitoring of cash balances, a company may not remain solvent for long. Cash revenues need to exceed cash expenses, and in recent years, A&E firms have seen their average collection periods for AR increasing.

You may not see the connection, but most of the topics we have discussed so far can affect cash flow. For example, if good Go/No-Go evaluations are not done, and time and money are spent on proposals for projects that the firm cannot win, or are not profitable, then cash is tied up in the proposal process and not free to be invested elsewhere in the business or used to cover payroll when collections are slow. Another example is that if timesheet management best practices are not followed, time may not be recorded accurately, and inaccurate bills could be sent to clients, which could slow down collections.

Chapter 2 detailed the length of time between the start of the project life cycle and when cash is first collected on a project. In some cases it could be a year or more. This requires you to have excellent cash management practices and backup financing to cover the business expenses when cash collected is not sufficient.

The main reason that cash flow suffers in most firms is slow collections of AR. When clients don't pay you, then you are financing their projects. As Chapter 2 explained, borrowing money to cover cash deficiencies can cost the company a lot of extra money in interest expense on borrowed funds. This ultimately has the effect of reducing project profit margins.

Best practices for improving cash flow include many of the areas that we have already discussed, but taking a look at them specifically from a cash flow standpoint can give you another reason to focus on improving these practices. In addition to the many tips and tricks I have learned over the years to improve collections, I reached out to some of my clients who have found that certain practices have served to keep cash flow in check in their firms.

The following business management and AR collections best practices successfully improve cash flow:

Improve timesheet accuracy – While I laid out many points above for why timesheet accuracy is critical, here are a few more from the standpoint of cash flow. Accuracy will dramatically improve cash flow by:

- Reducing the amount of time spent by administrative staff to correct incorrect entries
- Reducing PM approval time
- Reducing time to revise invoices
- Improving your chances of getting paid
- Eliminating the time it takes to deal with unhappy clients

Automate employee expense reports – This will allow you to bill expenses before processing and paying employee expense reports.

Set a fixed price for reimbursable expenses – If you charge clients for in-house expenses such as prints, use a fixed price per unit. This will reduces the amount of time to bill and eliminate many client questions.

Integrate with a cost recovery system – Some cost recovery systems allow you to track copies and prints, and automatically send the data to your accounting system for billing.

Have subcontractors bill clients directly, if possible – If you can write this into your client contracts, this may allow you to free up administrative time and keep you from having to manage pay-when-paid processes.

Bill clients twice a month – I have seen this work effectively for many companies. Some clients prefer getting more frequent, smaller invoices. This should speed up AR payments and improve cash flow overall.

Speed up the billing process – By reducing the amount of time it takes to get your bills out, you will speed up payment of your invoices by that many days. Some best practices for getting bills out faster include:

- Get timesheet and expense report approvals done weekly to minimize month end adjustments.
- E-mail electronic copies of all information that a manager needs to review their bill.
- Have a specified timeline for managers to return billing adjustments to accounting.
- Schedule meetings with managers to review their bills and get their edits.
- Provide as much information to the manager as possible to minimize time spent looking for project information, including billed and spent to date, AR information, percentage complete, and backlog by phase. We have designed a report called the Billing Review (for Deltek Vision) that has every piece of data a manager would need to decide what to bill. This can save them time and shorten the review process.
- Require consultants and subcontractors to bill you one week before the end of the month.

Make sure Invoices are easy to read and have all the required information – Here are a few tips that can help reduce billing questions from clients:

- Make sure the amount due is on the bottom of the first page. AP clerks often mistakenly enter the wrong amount into their system. The amount due should always be on the first page and preferably on the bottom right of the page.

- Do not provide more information than necessary.

- Add a statement to the invoice that the client must contact your billing person in writing within 30 days or the invoice is considered accepted.

- Bill expenses on a separate invoice or project so clients cannot hold up a large amount of services for a few expense questions.

- Negotiate with clients about having to provide backup with invoices. You may not be able to do this with government clients, but if you can, this will save your firm a great deal of time and money, and potentially reduce the billing cycle by days.

If you are sending backup documents, don't assume that you still have to. I once worked with an architectural firm that was sending backup to every client on every invoice. I asked them to send a letter to the clients telling them that they were "going green" and storing all receipt copies online in case they were needed. Only one of the firm's clients required them to keep sending the backup.

Develop Relationships with Clients' Accounting Staff – By having good communication with your clients' accounting staff you will find it easier to push invoices through their process. It is important to understand the challenges they have getting approvals for payment internally so you can better navigate their systems and processes. June Pride, Senior Accountant at McCormick Taylor in Philadelphia, Pa., regularly uses this practice to get results for her firm:

In the A/E industry, collections are more effective if you build solid relationships with clients from both a technical and financial perspective. It is certainly a case of you get more bees with honey.

Review the invoice with the client the first time you send it to them – Make sure they are happy with the format and information provided. This is especially important for new clients and government clients. Ted Maziejka, consultant to the A&E industry, takes it a step further. He has the following recommendation, which I think is great advice:

The first key to success is creating your project kick-off meeting to include your accounting staff so they meet the client accounting staff and have all the details worked out before you submit your first invoice. If your accounting staff cannot attend the meeting physically, allow for there to be an agenda item that clearly indicates to the client that these items need to be addressed and request the name and contact information of the client's accounting staff. All the details of the invoice format, timing of receipt for proper turnaround, and questions are sorted out during this review.

Send invoices by e-mail and snail mail – This helps eliminate the excuse of not having received the invoice.

Charge interest on late payment of invoices – While interest charges are often difficult to collect, they do have an effect on the client's payment choices. In some cases, you may collect them, and this is another bonus. Make sure this is clearly spelled out in your contract and agreed to upon in advance.

Call clients five days after the bill is sent to confirm they received it and that they don't have questions – This is an extremely effective method for getting paid faster. This will eliminate many collection calls and excuses about not receiving invoices or having questions that are holding up payment.

Call all clients 1-2 days after the due date to ask about expected payment – Don't wait until invoices are 45 days old to follow up on past due balances.

Record notes about collection conversations to assist in the event of legal collection efforts.

Use alerts to warn PMs and executives about non-paying clients – This will ensure that work does not continue when clients are not paying.

E-mail statements and/or send statements on brightly colored paper.

The squeaky wheel gets the oil! Call often – I did the books for an architectural firm for 10 years in the 90s. I can tell you that when the subcontractors called about payment, they were more likely to get paid faster. It does pay to be persistent and to call often.

If you are a subcontractor subject to pay-when-paid contingency with your client, try to write into the contract a process for resolving excessive payment situations, including your ability to go directly to the client to inquire about payment status.

In addition to all of these ideas, Bob Johansen, Principal, Leo A. Daly, gives the following innovative advice:

• •

Ever thought about what it would take to cut 30 days from your "total days outstanding" metric? The answer is simple. For all of your lump sum jobs, consider billing in advance. Before you dismiss this idea under the excuse that your client will never go for it, read on. Done correctly, your client will not even know what you are doing. Here is the old way of billing (billing in arrears):

1. *Earn revenue at the end of the month*

2. *Bill the earned revenue at the beginning of the following month*

3. *Wait to receive payment (ideally 30 days after generating the bill)*

4. *If you bill on the 10th of the month, your total days calculated = 10 unbilled days + 30 [AR] days = 40 total days. Sound familiar?*

To gain the upper hand, here is what the new process looks like (billing in advance):

1. *At the beginning of the month, instead of billing last month's earned revenue, estimate what the percent complete will be at the end of the month. You could use burn rate, future ETC hours, or input from the PM to do this. It is not that hard. Most projects are going to earn in a predictable way.*

2. *Create a bill based on your estimated percent complete and send it to the client. By the time the client receives the bill, reviews it, and approves it, the job should be close to your percent complete estimate.*

3. *Earn revenue at the end of the month.*

4. *What has changed? You have just shaved 30 days from your total days calculation! If you bill on the 10th of the month, your new total days calculation = -20 unbilled days + 30 [AR] days = 10 total days. In fact, my experience with doing this is that some clients will remit payment even BEFORE you earn revenue, making the total days for that invoice 0. This secret will supercharge your cash flow bring in the money!*

• •

Use the collection call as an opportunity to check-in with the client – Would you like to know why your client is not paying you? Ed Friedrichs, former CEO of Gensler, uses the collection call as an opportunity to check in with the client and build on the client relationship:

* *

On collections, I would add a caveat about how to make the call to a client about a past-due invoice. I always treated that call as an opportunity to get a "report card". Rather than thinking of this as a difficult or painful call to make, I used it to say, "If you're not paying our bill, we must be doing something wrong." Most of the time, the client was very happy with what we were doing and this served as that "squeaking wheel" call. When something was wrong, it gave me the opportunity to correct it fast, so it wouldn't fester. A problem not addressed, or worse, ignored or fought over, not only decimates cash flow but makes every other aspect of the project more difficult.

* *

ACCOUNTING PROCESSES FOR CASH

It is critical to put adequate cash management practices in place to eliminate the possibility of fraud or error. I have had at least seven situations in the last 23 years where my clients caught employees stealing substantial amounts of money. The types of fraud ranged from an employee forging signatures on checks to themselves or fake vendors, to charging personal items to company credit cards, to paying themselves extra in payroll. In some cases, these had been extremely trusted long-time employees, and the theft had been going on for years!

In addition to a background check on accounting staff, I recommend that internal controls are implemented to ensure that fraud cannot be executed. Good internal control processes include having a separation of duties between the person entering the transactions, signing checks, posting the transactions in the accounting system, and doing the bank reconciliation. Also, the checks should be reviewed to make sure they are being sent to valid vendors and endorsed by the actual vendor. Having an outside CPA or third party reviewing the financial statements and payroll journals is also a good idea. I am not suggesting that most employees cannot be trusted, it is just good business practice to put these controls in place to ensure the continuity and solvency of the firm.

SUBCONSULTANT/SUBCONTRACTOR MANAGEMENT

For many A&E firms, subcontracts account for up to half of the total fees on the project. Because of the substantial amount of work done by the subs, and the financial responsibility to collect and pay them when paid, most firms need to put in place controls to ensure subs are performing as required. This includes making sure they are following the scope of the contract, delivering the quality that is expected, billing according to the agreed process and milestones, and billing you the right amounts.

As a prime contractor, it can be a lot of work to manage everything being done by the subs. Some subs are small firms and don't have good accounting processes and systems. For that reason, it is especially important to review their invoices to make sure that you are being billed the correct amounts. While working in the back office of an architectural firm for years, I saw many engineers sending duplicate invoices for the wrong projects and incorrect percent complete. It was a lot of work to manage the processing and payment of the subcontractor invoices.

One recommendation is to use a PO system. This can allow you to record the contract value for each project that a specific vendor or subconsultant has, and more easily monitor how much has been billed against these amounts and is still remaining. It can also record committed amounts against the project to get more accurate visibility into net backlog and net fees.

Some systems also allow you to pay subs on a pay-when-paid basis. This allows your accounting department to track the sub's invoice against the AR invoice and determine when to pay the sub based on when you are paid. This can help to alleviate a lot of the hassle around managing subconsultant payments, and make it easier to respond to vendor inquiries.

If you are the sub, make sure your invoices or correct. If they are not, you are giving your client a good reason to delay payment to you.

MANAGING OVERHEAD

Overhead expenses are the expenses of the company used to run the business and are not directly attributable to a specific project. Common overhead expenses include indirect and administrative salaries, supplies, rent, utilities, fringe benefits, insurance, accounting and legal fees, and taxes. In a firm that has multiple offices or departments, the overhead may be collected in a corporate overhead department, and allocated out to all of the revenue-generating departments based on a formula. For example, rent may be allocated out to the mechanical, electrical, and plumbing departments in a MEP firm based on the headcount or square footage used by the staff in each department.

The basic overhead rate is calculated by dividing the overhead expenses by direct project labor. This tells you how much overhead needs to be applied to every hour of direct labor to cover the company's overhead costs. Figure 5 is an example of how an overhead rate is derived.

Overhead Expenses	
Indirect labor	$175,000
Employee benefits	$75,000
Rent	$500,000
Supplies	$10,000
Accounting and legal	$32,000
Total overhead expenses	**$792,000**
Total Direct Labor	$585,000
Overhead Rate $792,000 ÷ 585,000 = 135.4%	

Figure 5

120

Chapter 5 examined how billing rates are calculated and how unexpected increases in overhead rates can reduce or negate expected project profits. This can happen on any contract type.

Because of the impact that overhead has on project and firm profitability, it is critical to manage overhead carefully, and to put processes in place to control and approve purchases. But by far, the largest component of overhead expenses is indirect labor, which increases proportionately as utilization decreases. By managing utilization (the rate that billable resources are charging to direct projects versus their total hours), overhead will be lower, and project profits will be in control.

If your firm has federal cost plus government contracts, overhead is calculated and bid on the contracts based on the prior year rates. The government gives your firm a provisional rate to use for billing on invoices. At the end of each year, the billing amount is reconciled against your actual costs, and if your actual overhead rate is higher than your provisional rate you may or may not be able to claim additional costs at the end of the year. If the actual year-to-date overhead rate is lower than the provisional rate, you may have to give money back!

The government also has certain types of expenses such as advertising, interest expense, key man life premiums, penalties, and entertainment that may be unallowable for inclusion in the calculation of the overhead rate to be billed to the government. Government contract margins range from 6% to 12% of your cost, so unallowable expenses further lower your potential profits.

It is a good business management practice to put in place a corporate revenue and expense budget each year, and compare actual revenue and expenses to the budget on a monthly basis. By managing overhead expenses and rates closely each month, you will be forced to watch utilization and employee workload more effectively. This will ensure better project profit margins, as well as overall firm profitability.

Human Resources (HR) Management

HR management is possibly the most complex and difficult part of running a business. The best advice I can give is that HR rules are different in every state and can be much more restrictive than would even seem logical. Employment laws are extremely binding and confusing. Having both an HR expert to consult on hiring, firing, and any other employee issues and an employment attorney is key. A good employment attorney can help you write your employee agreements in such a way that will minimize your risk during an employee separation. Many documents, such as non-compete agreements will not hold up in certain states, and there are ways to protect the firm (although not completely in most cases) in the event of a dispute.

In addition to an experienced HR expert and employment attorney, it is critical to have someone on your staff who understands the payroll rules. Many times, I have seen my clients make costly mistakes that caused extended battles with the IRS and excessive penalties that could have been avoided. If you don't have a payroll expert on your staff, then your outside CPA firm may be able to provide this service. Even if you outsource payroll, it is still imperative that you have someone who knows the federal and state(s) payroll rules.

• •

Not following the rules can be very costly and risky. One of our clients discovered this the (extremely) hard way. The accounting staff was not knowledgeable about payroll rules and processes, and the company's outside IT consultants failed to install the annual payroll tax updates. As a result, payroll taxes were under-withheld for employees for a couple years, and the IRS assessed severe tax penalties and interest charges against the firm. An attorney was hired to help get the charges abated. In the end, the firm ended up spending more than $100,000 because of this mistake.

• •

I believe that most firms should outsource payroll. If all of your employees are in one state and you have an experienced person on your staff, then it might make sense to do your payroll in-house. But if you have employees that reside in more than one state, the level of difficulty of compliance starts to increase. If you have employees in three or more states, then you should definitely consider outsourcing payroll. The risks are just not worth the costs of outsourcing.

One common mistake is having 1099 consultants that should be treated and paid as employees. The IRS has strict rules about how to classify an employee and a 1099 vendor. George E. Christodoulo, PC, LAWSON & WEITZEN, LLP, Boston, Mass., warns about the increasing risk of misclassifying workers:

The monitoring and investigation of labor practices by state and federal agencies has become more proactive in recent years. Specifically, two focal points are targeted: (1) the misclassification of employees as independent contractors, and (2) the misclassification of hourly employees as exempt employees. Best practices include auditing the classification of all independent contractors and employees to ensure that your firm will not be subject to claims by the individuals involved and/or claims by agencies for payment of withholding taxes and overtime compensation.

There are many other HR topics that are not being covered in this book including recruiting, hiring best practices, employee retention, and compensation. These topics could be an entire book on their own and are well-covered in many other books. This can be a difficult topic for many non-experts, and if you are not proficient in this area, I recommend that you get the assistance of someone who is.

MEASURING FIRM PERFORMANCE

Measuring the performance of an A&E firm on a regular basis is critical to success. Yet many firm executives tell me that they are not sure how to get the information that they need in order to determine if their employees are performing as needed to return the profits they expect. As we discussed in Chapter 3, strategic planning, including setting goals and objectives for the company, is critical to success. The next step is establishing the company budget, which provides the overall guidelines with which to measure how the firm is performing on a monthly basis.

Financial Statements

Once the budget is done, the monthly accounting process should be completed and financial statements produced. The two primary financial statements include the balance sheet and the income statement, also called the Profit and Loss (P&L).

The balance sheet shows the condition of the firm at a given point in time. It shows the assets, liabilities and owners or shareholder's equity in the firm on a specific date, usually the end of a month or year. I like to describe the balance sheet as showing who owns the assets of the company—the creditors or the owners. If a company has borrowed or lost a lot of money, the creditors may own a greater portion of the assets than the owners.

The income statement shows the company's performance over a period of time, such as a month or year. It shows the revenue of the company, calculated as the billable value of work (both billed and unbilled) on an accrual basis, and cash received on a cash basis. After revenue is totaled, the direct expenses of the firm are subtracted to calculate the gross margin. Direct expenses are the expenses directly related to creating the revenue, including direct labor and project direct or reimbursable expenses. Finally, the overhead expenses are listed, including fringe benefits, indirect labor, and all of the other expenses required to run the business. The amount left is net profit.

The income statement can be presented many different ways depending on the organizational structure of the firm (office, departments, etc.), and whether actual costs are being compared to the budgets. Another beneficial way to look at the income statement is to use a trend analysis, which shows how revenue, expenses, and profits are trending over a period of time.

Looking Back: Key Performance Indicators (KPIs)

Chapter 2 detailed where to find the lost dollars in your business. This is a good first step in determining the metrics that need to be tracked and reported. Your reporting rhythm will be determined by your reporting systems and processes, as well as what is required to ensure that you have adequate visibility into the important aspects that need to be reviewed.

It is valuable to compare your metrics to industry standards that are available through organizations such as ZweigWhite or PSMJ. You should also look at the monthly trends of each metric over time to highlight problems or other issues affecting the consistency of your performance. Depending on your business model, the following are metrics that I recommend you look at on a monthly basis:

Win Rate = Number proposals won ÷ number proposals submitted. This shows the firm's success in winning projects on which you are bidding.

Net Revenues = Total Firm Revenues - Subcontractors and other direct costs. This is the amount of revenue attributable to the services of the firm only. This can be calculated for the firm as a whole, or averages for projects or employees. Many firms look at Net Revenues per employee as a measure of success.

Frank Stasiowski, FAIA, Founder of PSMJ, recommends looking at a rolling net revenue per employee as an effective measure of firm performance: *"I am a big believer in trends. Firm leaders really need to measure six-month and six-year rolling trends for key metrics – especially net revenue per employee. This will tell you a lot more than just looking back at the current year."*

Project Profit Margin = Total Project Fees - Total spent on Project (including overhead allocation) ÷ Total Project Fees. Project profit margins are expressed as a percentage. Many firms calculate the project profit margin as a whole for the firm, as well as for individual projects, clients, departments, project types, and PMs. This can help you understand where problems are occurring, and which managers are most effective.

Utilization Rate = Direct hours ÷ Standard hours (usually 2,080 per employee, but sometimes holidays and PTO are subtracted). Utilization should be reviewed each pay period or at least monthly by employee and department, and for the firm as a whole. Employees should each have a target rate and be able to self-monitor their own utilization. It can also be calculated using dollars, and you will get a different result depending on whether your more or less expensive employees are more billable.

Realization Rate = Direct hours billed ÷ Direct hours worked. This rate shows whether the employee's billable hours are able to be billed. This is sometimes a better measure of employee productivity than the utilization rate, especially if your firm frequently holds time or allows write-offs on T&M contracts. For fixed fee contracts, realization is often measured by taking the total amount billed on a project and allocating it to each employee by the relative billable value of the time charged. This rate is used to measure the firm's ability to bill and collect for work performed, and low realization rates often indicate issues with billing practices.

Overhead Rate = Total overhead expenses ÷ Direct Labor. Expressed as a percentage, this is the rate that needs to be applied to each dollar of direct labor by project to calculate the true project profitability.

Net Multiplier = Net Service Revenue ÷ Direct Labor. This shows you the multiplier against each dollar of direct labor that is being recovered in revenue. Firms have traditionally operated with an average of a 3.0 multiplier (setting the billing rate three times the raw labor cost of an employee).

Average Collection Period = Total Accounts Receivable ÷ (Gross Revenue ÷ 365). This is the number of days on average that it takes to get paid on your invoices.

Firm Profit Margin = This is the overall profit of the company as reported on the income statement. There are many different calculations for the firm profit margin that either include or exclude certain expenses such as bonuses, interest, taxes, or other unusual expense items.

In addition to these key metrics, you should be looking at some standard accounting metrics such as:

Current Ratio = Current Assets ÷ Current Liabilities. This metric shows how solvent the company is at a point in time. It indicates the company's ability to cover short-term debt requirements with cash and cash equivalents.

Debt to Equity Ratio – This metric shows how leveraged the firm is in debt. Too much debt can be dangerous. Sometimes too little debt can hurt the firm as well, leaving it undercapitalized.

Return on Owner's Equity = (Net Profit before bonus and taxes ÷ Owner's Equity) x 100. This shows the ROI for the owners.

If the firm is planning to be acquired at some time in the short- or long-term future, maintaining books according to GAAP is critical (see below). Jonathan Voelkel, Director of Special Projects at Rusk O'Brien Gido + Partners (ROG) in Falls Church, Va., writes about the importance of good accounting practices for a firm that is being evaluated as an acquisition target:

In the stage of courtship when an acquirer gains access to a target firm's financial statements, the quality of reporting can have a significant effect on how to further pursue the opportunity. At best, poorly presented financial statements can be a big headache for both sides, and can often lead a seller to incur additional costs related to providing satisfactory reporting. At worst, the integrity of an organization can be called into question when transparent statements aren't available.

• •

Looking Forward: Forecasting

It is great to look back and see how your firm is performing, but this does not necessarily indicate your future profitability. In order to try and forecast your firm's future success, you need to be able to look at three other key metrics or figures that will give you a much better indication of how the firm will perform in the following three to six months:

Backlog = Total signed contract value – total billed and spent (unbilled) to date. The backlog shows you how much signed contract value you have for the future. It can also give you an indication of how many more months of work there is, how your marketing efforts are performing, and whether you may need to hire or lay off staff.

Opportunity Pipeline – In addition to signed contract work, you also need to be able to estimate the amount of work that is coming in from current proposal efforts. Pipeline is the estimated amount of work that you believe will be awarded in the following six months. It is usually calculated by applying an estimated fee value to each proposal, as well as the probability that you will win the award, providing a weighted revenue calculation. Most pipeline reports show expected project awards by month.

Cash Flow Forecast – This is an analysis of all expected cash receipts, based on current AR and normal collection periods, less the expected cash commitments for normal business expenses, purchases, payroll, subcontractors, and debt financing. This analysis gives you a picture of whether the firm will be able to meet all cash requirements and an indication of whether you should potentially consider borrowing on a line of credit. This can be quite complex when also considering potential cash revenue from backlog and the pipeline.

CASH AND ACCRUAL ACCOUNTING

Most A&E firms are using some variation of both cash and accrual accounting. If you are filing taxes on a cash basis, then it is important to stay on top of how the firm is doing on a cash basis so that you can try to forecast your potential tax liability.

Each method of accounting gives you a different view of the performance of your firm. Cash accounting really only tells you how well you are collecting money, and if you are receiving more money than you are spending it. It does not show you how the company is performing based on the work that is being done in the current period. Cash receipts can come in for work that was done last year or last month and does not match the expenses that generated that revenue to the correct accounting period. You will not see AR or AP represented on cash basis financial statements.

In addition to cash financial statements, most firm owners are looking at (or should be looking at) accrual basis financial statements. This is where the revenue and expenses are matched in the correct accounting period and are based on dollars committed to spend and work performed, regardless if any cash is paid or received. An example is payroll expenses and billing revenue, where the payroll, even if paid to the employees in the following month or year, is charged to the period where the work is done, and the billing revenue, even if billed in the following month or year, is accounted for in the period that the work was done. In addition, other expenses that are paid for in a given accounting period must be spread evenly over all the periods that the expense provides benefit, such as insurance, prepaid rent, or bonuses. This ensures that an accurate picture of the firm's performance is represented on the financial statements.

True accrual accounting is dictated by GAAP. What I have seen in most smaller, privately owned A&E firms is what I call modified accrual accounting. This is where AR and AP are represented in the correct accounting period, but other expenses are not accrued and allocated to all the periods where they provide value. I have also seen practices where firms are posting revenue and payroll to the periods in which they are paid, rather than when they were incurred. I call this modified accrual because the financial statements have AR and AP on them but do not properly match revenue to expenses.

Accrual accounting gives you a much better picture of how your firm is performing, and allows more consistency in analyzing trends from month to month and year to year. However, it is still possible to have a great deal of income on an accrual basis, and the firm still goes out of business because of failure to collect what has been billed. For that reason, it is important to look at both accrual and cash financials statements to get an accurate picture of the company's performance and health.

A big issue that can impact how you prepare your financial statements is whether you have outside shareholders, or a bank, that require GAAP financials on a monthly basis. In this case, you should have a CPA on staff, or at least access to a CPA to assist in correctly recording financial transactions and presenting the financial statements on a GAAP basis.

GOVERNANCE

Managing risk is a critical aspect of running a successful and profitable business these days. It is the responsibility of the owners of your firm to ensure that all legal, regulatory, and tax rules and processes are followed and managed. This will require the hiring of a good CPA firm and law firm. I am a big believer that money spent up-front to prevent potential risks is not only good advice, but it is also critical.

One of the areas where this is most true is in your employment, business formation, and client legal agreements. By adequately protecting your rights in the event of a dispute, you are potentially saving yourself huge legal expenses and possible losses on the back end. While I cannot comment on all the potential areas this could be relevant, I would use this as a rule of thumb in general for all legal agreements.

Another area of significant risk that should be confronted is the compliance aspect of managing the financial, HR, and operational aspects of your day-to-day work. Too often I have seen clients get in trouble by not ensuring that their business processes are supporting the compliance with all government-mandated regulations. An example of this would be to ensure that terminations are done with a certain process, along with adherence to the recommendations of a knowledgeable HR professional. With the number of lawsuits these days from disgruntled and terminated employees, it is prudent to ensure that your performance review and documentation processes are iron clad and support all of your hiring and firing decisions.

If your firm is too small to have in-house experts in all of these areas, it is advisable to outsource these services. It is critical to balance the cost of these services with the potential foreseeable risks and to find the balance of cost versus potential losses.

Board of Directors (BOD)

In addition to having the services of legal, accounting, and HR professionals available to you for legal, tax, and government compliance matters, you may want to assemble an Advisory Board or Board of Directors. Wikipedia defines a BOD as "a body of elected or appointed members who jointly oversee the activities of a company or organization." (Source: http://en.wikipedia.org/wiki/Board_of_directors.)

A BOD can be a valuable resource for helping to determine the firm's strategic direction, making critical business decisions, and ensuring another layer of governance. By gaining access to industry and other related professionals, you will be able to get a different perspective on the firm's leadership, and another more objective view on the SWOT of the company (see Chapter 3).

Members of a BOD are paid a stipend and expenses, not paid at all, or offered some kind of stock or stock options in exchange for their time and expertise. The role of the BOD can vary between firms, but a typical BOD will usually have the following role:

- Governing the organization by establishing broad policies and objectives
- Selecting, appointing, supporting, and reviewing the performance of the chief executive
- Ensuring the availability of adequate financial resources

- Approving annual budgets
- Accounting to the stakeholders for the organization's performance
- Setting the salaries and compensation of company management

There can also be challenges with governance when there are different ownership structures. It is critical that decisions are made in the best interest of the firm with as little disagreement and politics as possible. Ed Friedrichs, former CEO of Gensler and author of "*Long-cycle Strategies for a Short-cycle World*," succinctly describes the challenge as a firm grows:

• •

Governance embodies leadership and decision responsibility. Every firm needs to identify where guidance and direction comes from and where the buck stops when tough issues are confronted. It's easy when a firm forms around an individual. Partners add more complexity. They're never perfectly aligned so they must come to terms with each other's differences to avoid confusion for the firm. Corporate structures in larger firms add another level of potential for conflict over authority and direction. And, of course, generational succession requires the firm's leadership to come to terms with new personalities, and often, values, as governance transition takes place.

• •

The core rule of thumb for ensuring an appropriate level of governance is to cover your risk on the front end. Preventive measures are much less expensive than dealing with problems and law suits later on. Paying lawyers to write solid agreements is much less expensive than paying them on the back end to defend a lawsuit or charges by government officials.

In addition to protecting the firm from potential legal problems, you should also consider ethical training for your management and staff.

With more bad press in the news these days about ethical violations or indiscretions, it is important to educate your team about the company's values and emphasize the need for integrity in all business operations.

Summary

- Most principals of A&E firms have never had formal financial management training. It is essential to have a financial advisor who is trained to manage the finances of the business. A good financial expert can help save the company money by making better decisions, avoiding taxes and penalties, and managing overhead costs.

- Cash flow management is critical to the survival and growth of an A&E firm. Improving many of the business processes and systems discussed throughout the book will also have a positive effect on cash flow.

- We outlined many best practices that are critical to both firm management and cash flow. The benefits of these processes go beyond having sufficient cash for operations, including increased accuracy, more efficient billing and collections, and higher client satisfaction.

- It is critical to implement effective internal controls to ensure that cash is protected from fraud, embezzlement, and error.

- Subconsultants can be a very significant portion of an A&E firm's contracts and should be managed to ensure quality, accuracy of billing and synergy with the project team.

- Overhead expenses include the costs of running the business that are not directly charged to a client project. Overhead should be allocated to projects, usually proportionately by direct labor charges each period. If actual overhead expense rates are higher than the rates used to develop the firm's billing rates, the projects will not make as much money as planned, or possibly lose money. Overhead expenses should be budgeted each period and managed to ensure that they are under control and the actual overhead rate is close to the targeted rate.

- The area of HR requires a great deal of knowledge of the law. Defined processes should be developed for every aspect of hiring and managing employees. Because of the complexity of HR rules, and the differences of laws in every state, you should have an HR advisor available to consult with for many employee activities.

- Recommended monthly reporting includes financial statements such as the balance sheet and income statement, as well as the calculation of KPI's. It is also important to forecast future pipeline, backlog, revenue, and labor requirements in order to ensure sound marketing and staffing decisions.

- Every firm needs to have a system of governance to ensure that the company is being operated soundly and guided with effective management decisions. A board of directors or advisory board can be an additional decision-making resource to provide another level of scrutiny and analysis to ensure appropriate guidance for the firm.

CHAPTER - 8

MANAGING CLIENT RELATIONSHIPS

> *"It's not the employer who pays the wages. Employers only handle the money. It's the customer who pays the wages."*
>
> ~HENRY FORD

One of the things that is universally important to A&E firm owners and executives is protecting and improving their relationships with their clients. With increased competition for projects, it is more important than ever to ensure that client relationships are managed and nurtured. Jay Appleton, Principal, Kitchen & Associates, Collingswood, N.J., explains the challenge caused by the recent economic downturn:

In this new economy, there is a pervasive threat - that of (new) competitors that are infiltrating every market possible, regardless of qualifications or experience, and competing on the basis of unrealistically low fees. When exacerbated by an owner's procurement philosophy focused on cost rather than value, this dynamic can result in tremendously wasteful resource burn and unacceptably low return on marketing investment related to costly opportunity pursuits. The lesson learned, after several bad experiences, is that steadfast reliance on a culture of prudent and thoughtful client care, and relationship-based business development, leads to productive, profitable work, and desirable long-term relationships.

• •

Technology that helps with managing client relationships can be a big advantage for A&E firms. The purpose of a CRM system is to help you manage both prospective and existing client relationships by improving communication, recording relevant data and activities, and scheduling follow-up calls and meetings. A good system can make it easier to send reminders and notifications about things that are going on with clients that may need attention.

A CRM system can also help you remember important deadlines, follow-ups, and promises made to clients. My client, Doug Hansford, President of Ben Dyer Associates in Mitchellville, Md., laughingly calls CRM "can't remember much." This is a funny play-on-words but is actually rather close to the truth. The CRM remembers important events and scheduled activities so you don't have to. I like to say that it helps get rid of a lot of sticky notes and e-mails.

Because a project is the center of an A&E firm's universe, tracking correspondence, communications and deadlines at the project level is important. Getting your employees to follow established processes for recording this information is also important but can be difficult. The following is a list of many of the benefits of a good CRM system that, when implemented and used effectively, can provide a competitive edge:

- Consolidation of multiple sources of client and contact data
- Enables implementation of automated marketing and sales processes
- Allows tracking and analysis of marketing efforts and results
- Facilitates better pipeline management and forecasting to increase win rates
- Permits better insight into client satisfaction and project progress
- Lessens the impact of employee turnover
- Automates the "to do" list for faster and easier follow-up
- Creates a higher level of accountability of employees
- Lets you know if you are keeping your promises to clients
- Reduces the number of emails and sticky notes at each person's desk
- Allows easier automation of a Go/No-Go process
- Automates client e-mail processes (e.g., holiday list and mailing, etc.)
- Notifies users about certain events or changes to data in the system
- Enables integration with Microsoft Outlook or other e-mail systems
- Facilitates automation of client loyalty programs
- Helps keep deadlines, proposal due dates, and other follow-up activities from falling through the cracks
- Gives executives more visibility into what is happening with clients and projects

To determine the ROI from a CRM system, you should evaluate how your key metrics will be affected by gaining the benefits listed above. As we reviewed in Chapter 2, by improving the way your firm selects projects to bid on and manages opportunities, you can improve your win rate and significantly decrease the cost of pursuing work. This will in turn also have the benefit of better project success because the projects being won are more appropriate and a better fit for your firm. In addition to the win rate, the cost savings from increased client service and retention can be dramatic.

It is well-known that it can be very difficult to implement a CRM system. Many CRM system implementations fail for a number of reasons. The major reason is user adoption, an inability to get employees to follow subscribed processes and consistently manage data. With careful planning and a clear concept of what the end goals are for the CRM, you can have a successful CRM implementation that will make a significant difference in your business operations and success. Chapter 9 details how to select and implement software systems and best practices to ensure appropriate levels of user adoption from your staff.

In addition to implementing a CRM, another effective practice that will help you remain in good standing with your clients is to ask for feedback on a regular basis. Surveying clients both during and after your project is completed will provide invaluable information and will allow you to deal with potential issues before they become full-blown problems. There are software systems that can help with this process, making it easier to send surveys and analyze the data.

Tom Peters, best-selling business book author for more than 30 years, discusses the relationship between measurement of customer satisfaction, and business success:

● ●

Regular, quantitative measurement of customer satisfaction provides a much better lead indicator of future organizational health than does profitability or market share change. We suggest monthly measurement. Further, we urge participants to make the level of customer satisfaction the primary basis for incentive compensation and annual performance evaluation for virtually every person at every level in every function throughout the organization. (Source: What Gets Measured Gets Done, April 2006, http://www.tompeters.com/column/1986/005143.php)

● ●

What you do with this valuable information is critical. David A. Stone, CEO of Stone and Company in Everett, Wash., is a consultant to A&E firms in the area of strategy and business development. He warns that the follow-up to asking for feedback is just as important as getting the feedback itself:

The first lesson in looking for client feedback is a word of caution. Regardless of how you learn what clients think, once they've shared their thoughts they fully expect you're going to do something about it. If you're not prepared to make real changes based on the feedback you receive, don't ask for it. Soliciting opinions and then doing nothing is far worse than failing to ask in the first place. If they had some concerns before, they're downright pissed now.

While software can help a lot to ensure improved client relations, there is no substitute for good old-fashioned face-to-face meetings. Having a focus on nurturing important client relationships is a strategy embraced by the most successful firms, and has one of the highest paybacks of all types of marketing and business development activities in which you can engage. Your CRM can help you remain at the forefront of this process by automating reminders and other notifications that can help you stay on top of client events, important dates, and quarterly check-ins.

Ron Worth, CEO of SMPS in Alexandria, Va., and author of *Building Profits in the AEC Industry,* has manygreat suggestions for improving your firm's client relationship practices. His recent article written for our Acuity Business blog gives 10 rules for networking to improve both client relationships and business development efforts in general:

Make customers and clients your friends—networking is an exceptional way to expand your circle of friends, support, and resources. The friend aspect of networking is by far the most rewarding and fun. Each new contact has the potential of developing into a lifetime friendship. And 9 times out of 10, these contacts are the ones who deliver your next contract.

• •

For the complete article on this topic, go to http://acuitybusiness. com/blog/bid/243726/Networking-10-Rules-to-Jumpstart-Your-Career-and-Get-Your-Team-on-First-Base.

As mentioned in Chapter 1, categorizing your clients into "A," "B," and "C" (and "F" for some) you can start to see which clients provide your biggest revenue and profits and begin taking a more strategic approach to your client relationships.

Obviously revenue is not the only criteria to consider when deciding how to categorize your clients. Other considerations might include how many projects you have done with them, whether they are a reference, payment history, referrals, and even subjective criteria such as how much your team enjoys working with them or how easy they are to work with. Once the clients have been categorized, you can develop an outreach and loyalty plan that rewards your "star" clients, and ensures that the relationship gets the attention it deserves.

Meeting with your clients, getting feedback, and making sure they have a stellar experience with your firm is the key to client retention and continued repeat business. John Russell, President of Harley Davidson, says it better than anyone else I have seen, *"The more you engage with customers the clearer things become and the easier it is to determine what you should be doing."*

I believe that by meeting with your best clients, and asking them about their specific challenges and goals, you can better help them achieve their goals, and form the strategy for your own firm's future. By helping your clients succeed in as many ways as possible, you will increase the value that you deliver to them, beyond the basic requirements of your contracts.

Summary

- Developing and managing excellent client relationships are critical to the success of your firm. It is a lot more expensive to find a new client than it is to retain an existing client.

- A good CRM system can help automate many of the processes involved in pursuing new business, as well as manage existing client interaction.

- CRM systems allow you to track correspondence and activities such as e-mails, meetings, and phone calls between your team and your client. This provides many benefits such as greater documentation and visibility across the firm on opportunities and projects, avoidance of missed deadlines, and better ability to keep promises made to clients.

- Asking for and following up on client feedback is an effective way to stay on top of client satisfaction. By categorizing clients into "A," "B," "C," etc. you can begin putting programs in place to ensure that your best clients get the most attention. You can use many different criteria to determine how to categorize clients, including revenue, payment history, repeat business, and various subjective assessments.

- Meeting with clients in person is the best way to deepen relationships. These meetings can also help you assess how to assist your clients in new and different ways, which can, in turn, help expand your value to your clients beyond your traditional services.

SYSTEMS AND IT

"The architecture and engineering profession has always been a notoriously slow adopter of technology. Hopefully, that's not the case with your firm, but if it is, it gives you a great opportunity to move to the forefront of the A/E industry in adoption of new technologies in just a short amount of time.

Adopting a strategy of embracing new technological ideas will impact many aspects of your firm's operations. Of course, it means making investments in software, computers, and mobile devices, but the strategy is much more than that. It impacts your leadership development, your hiring, your project management, and nearly every phase of firm operation."

~FRANK STASIOWSKI, FAIA, IMPACT 2020: 10 GIANT FORCES NOW COLLIDING TO SHAKE HOW WE PRACTICE DESIGN IN THE FUTURE, PSMJ RESOURCES, INC., 2010

Systems can make your life easier, and if implemented effectively, they can provide synergy, streamline processes, and improve operations. Technology has changed dramatically over the last few years, however many A&E firms are entrenched in software products that were effectively designed in the 80s, 90s, and if lucky, early 2000s.

Changes in technology have forced software publishers to rewrite their systems, often several times over the lifetime of their products. Business management software is the best example of this situation. I have followed the market closely for more than 23 years, watching software companies come, go, and get acquired, forcing their users to phase out older legacy systems and search for new software to run their front and back offices.

When I started working with A&E firms in 1989, the most popular accounting packages were Micromode, Databasics, ACCI, Wind2, and Harper & Shuman CFMS (later called Advantage). Recognize any of these names? If your company has been around since then, you probably had one or more of these systems and went through the pain of changing whenever one went out of business or was acquired.

My company supported Wind2 for more than 15 years and helped many A&E firms migrate from these other systems, only to later convert to Vision when Deltek acquired Wind2 in 2005. Some of our clients have switched four or more times in the last 20 years. There are now other options in the market as well, which makes the selection of the system even more difficult and risky.

As technology and software continues to improve, there are corresponding challenges in staying current and ensuring that it is embraced and adopted by your staff. User adoption is one of the most challenging aspects of technology investment for a firm. The worst thing that can happen is for the firm to invest in technology with a goal of improving the business management and profitability, only for the implementation to fail or groups of users in the company then fail to use the system as planned.

Here are just a few of the challenges faced by today's firms that are only just starting to be addressed and overcome:

THE CHANGING WORLD OF TECHNOLOGY TODAY

Bring Your Own Device (BYOD) – As our workforce becomes more mobile and our employees bring their own devices in to the office, we have several challenges. First, we must realize that we can't control what they are doing on their own phones or tablets. Second, we must figure out how to secure our company systems from viruses and other threats that can be more easily introduced into our environment.

Social Media – The way that people interact these days is very different than how it was just a few years ago. Social media can be a challenge to firms in managing their brand images online, controlling news and confidential information, and understanding how our employees are interacting with clients, prospective clients, and others with whom we do business. It is also a challenge to control the image that our employees project individually as they mix personal and business contacts in a single platform like Twitter or Facebook.

Inbound marketing – As explained in Chapter 4, inbound marketing can be a great competitive edge, but requires a great investment of time and resources to be successful. Often the ROI from inbound marketing takes years to recognize. To be successful at inbound marketing requires a great deal of high-quality content delivered to clients and the "industry" on a weekly basis. Patience is critical while building a following online, and you may need to take a long term approach to measuring success in this area.

Search Engine Optimization (SEO) – In order to be found on the Web, you must invest in SEO. This is the practice of constantly monitoring the keywords that people are using to search for the services your firm offers, and "optimizing" your web site and content to improve the chances you will show up on the first page of search engines (Google, Bing, Yahoo, etc.). SEO is not something you can hire someone to do and then let it go for a year. The top search engines are constantly changing the criteria used to determine how your Web site shows up in the listings, and it requires a great deal of expertise and regular consistent monitoring and adjusting in order to ensure your firm remains competitive.

145

Cloud Solutions – The trend in the industry is for more data, communication, and applications to reside on the Internet, also called the *cloud*. The cloud represents any format of taking a software application and data, and implementing it so that it is available online. This can be done by putting your server in rack space in a data center, finding a hosting provider, or by using a SaaS application (see below).

The benefits of the cloud are obviously the access to the application and your data from any location, and the advantage of not having to buy servers, backup devices, security, and IT support. For many companies, it is a no-brainer to send their business management software into cyberspace. One of our clients, Larry G. Kirk, CPA, Vice-President of Finance at AES Consulting Engineers headquartered in Williamsburg, Va., is very happy with his decision to move his Deltek Vision system to the cloud:

• •

One of the operational decisions that has most affected our business is moving to hosted solutions for some of our IT needs. It has allowed us to get back to the business of running a business instead of running our IT. It is one of the best decisions we have ever made.

• •

Software as a Service (SaaS) – As the cloud becomes more popular for data storage, most software publishers are moving their systems to the cloud and offering SaaS. This is a completely new business model, both for the software companies and their customers. It basically means that instead of buying the software and paying annual maintenance for updates and support, you will pay an annual subscription to effectively "lease" the software.

SaaS can be a great benefit to some firms that have had issues with keeping up with the latest versions of their software applications, hardware failure, backup problems, or consistent internal IT support. In a SaaS licensing agreement, the software publisher manages the IT component of the software, hosting the software on their servers, and providing normal IT support such as updates, backups and security. They make the software available over the Internet, allowing access from any location. Other benefits include reduced risk from power failure, employee sabotage, and mistakes by IT staff who are not experts in managing the application.

The main disadvantages are recognized if a firm has needs for customization, or integration to other applications. In most cases, this can be difficult if the provider does not have a sophisticated API that allows other systems to connect to the database.

Cyber Security – As much of our data is now housed on servers in our offices, or in the cloud, security is a critical factor in our IT management. There are threats from many areas, including online hackers, viruses and malware, and even employee embezzlement and sabotage. Every company needs to have a security plan that encompasses all of these areas. Attention needs to be given to systems and monitoring processes that protect the firm from known and unknown predators.

Managed/Outsourced IT Services – In addition to moving technology to the cloud, many firms are finding that outsourcing IT services is a more cost-effective and strategic way to manage the company's hardware and applications. This can be a good option for firms that have many offices, are too small for a full-time IT resources, or don't have enough work to keep a full-time person busy. In addition to typical IT services such as server management, updates, backups, and troubleshooting, outsourced IT services may also include remote monitoring of servers and desktops, virus protection, and security. Some of these services may not be required if you are using SaaS solutions.

ERP – As I mentioned earlier, ERP business management systems are becoming the most popular choices in professional services businesses today. Integrating all of the data in your firm will save time and money by eliminating inefficiencies and redundant manual process in your marketing, accounting, and project management processes. It can also make your firm more profitable. We will look at the benefits of an ERP system in further detail below.

Remote or Virtual (Home-based) Employees – There is a significant trend in the United States for employees to work at home or in small remote offices. While the Internet and changing technology makes this much easier to do than ever before, there are definite challenges in communication, marketing and proposal management, and execution of work on projects that must be addressed. The IT department of your firm must be able to work remotely to support the employees' remote systems, and processes must be put in place to allow remote workers to access critical company applications such as e-mail, documents, timesheets, and project reporting (depending on their role).

In addition to remote employees, more firms are hiring and opening offices overseas. This trend creates additional requirements for cloud-based applications, security, multi-currency conversion, and remote monitoring of systems.

In looking at the trends affecting the market at this time, it is almost impossible to stay on top of all the constant changes. The role of the IT guy is changing to be more of a strategic position, rather than a geek that does troubleshooting for the staff. He needs to be aware and educated about changes in technology, so that he can advise us to make the right decisions now. Considering that we are locked into most IT purchases for many years, looking at software purchases with a long-term strategic view can save your firm money down the road.

OTHER AREAS WHERE TECHNOLOGY CAN HELP YOU FIND THE LOST DOLLARS

Chapter 2 suggested taking a different look at your business and calculating the potential savings by improving the processes, inefficiencies, and systems in your business. As fast as technology is changing, you may never totally get caught up and have the latest and greatest systems. But by focusing on the metrics that have the largest impact on profits, and by making the changes we discussed in Chapters 3 through 8, you should be able to start seeing gains in less than six months.

Many areas of your business can benefit from improved technology and affect your overall success. The list below is a fairly comprehensive list, but with as fast as technology is changing, something new could come along soon that might make a difference as well. Some areas where technology can help you improve business management include:

- Market intelligence (government projects and RFP announcements)
- Marketing, content management, and lead generation
- Social media
- Estimating
- Accounting
- AutoCAD/Drafting
- Pipeline and opportunity management/forecasting
- Proposals
- Time and expense management
- Project management and budgeting
- Billing
- HR
- Resource management/scheduling
- Project/Client collaboration using web sites or portals
- E-mail
- Document management
- Cash flow and collections
- Banking/online bill pay and cash management
- Financial analysis/business intelligence
- AP automation/purchasing
- Meeting collaboration
- Client feedback

LOOKING AT TECHNOLOGY AS A STRATEGIC INVESTMENT

Obviously with all these areas of your business that can be automated, there could be significant costs involved in purchasing and implementing these solutions. There is a risk in trying to adopt new technology, as well as a cost and risk in continuing to use older, outdated systems. I believe that business management software should be evaluated and viewed as a strategic investment. Based on your goals for growth, improved efficiency, and higher profit margins, technology is the cheapest and best way to achieve these goals. If you compare the cost of software to what it costs to hire and retain human capital, you can see that software is a much less expensive investment.

In evaluating and purchasing technology, you need to be able to get a ROI from it. It is often difficult to understand the cost of that investment, as the total cost of ownership (TCO) of today's ERP systems are difficult to calculate. Many factors must be considered in trying to understand the total cost of technology including:

- Software licenses (up-front or annual subscription)
- Annual maintenance (if perpetual license)
- Implementation/consulting services
- Training
- Additional follow-on support
- Internal implementation costs for employees including accounting, marketing, IT, PMs
- Annual hosting costs (if deployed in a hosted environment)
- Power and Internet connections
- Additionally, if you decide to have the software reside in your office, you will also have:
- Hardware: servers, backup systems, redundancy, and firewalls
- Operating systems and other software: SQL or other database software, server operating system licenses, virus/malware protection, etc.

- Security and heating and cooling for servers
- Additional IT management support (backups, updates, server management, etc.)

An integrated ERP solution can provide a much higher ROI than separate non-integrated programs. As detailed in Chapter 2, by consolidating separate silos of data and non-integrated processes, you can realize cost savings by improving administrative efficiency, eliminating redundant processes and data entry, enhancing accuracy in time collection and billing, decreasing budget overruns, and maximizing resource utilization. By evaluating your current processes and where cost savings can be targeted, you can better evaluate each system to understand where it will be able to achieve the desired results.

Aberdeen's November 2011, *ERP in Professional Services: Managing Costs When People are the Product* report found that 73% of *leading* professional service organizations are using ERP compared with 60% of followers.

Professional Services Implementing ERP

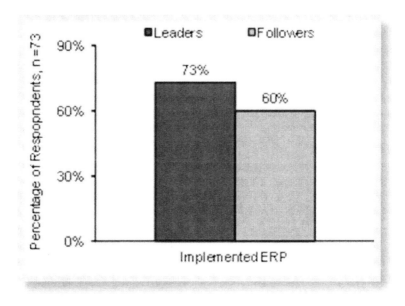

Figure 6

Aberdeen defines ERP as an integrated suite of modules that forms the operational and transactional system of record upon which any business is based. The visibility, standardization, and process efficiencies that are derived and enabled by ERP allow professional services firms to more effectively cut costs, manage projects, and staff accordingly. The results are more consistent services and satisfied customers. At the same time, 74% of leaders have real-time visibility into the status of all processes from quote to cash compared to 32% of followers. This facilitates revenue recognition for professional services organizations where it is more difficult to define when services have been delivered to the customer. (Source: Nick Castellina, Research Analyst, Enterprise Applications, AberdeenGroup)

As Aberdeen cited above, an ERP solution is an integrated set of software modules, or programs that help a company automate business processes throughout the company. For an A&E firm, this is usually a standard set of accounting modules, including GL, AR, AP, Timesheets and Expense Reports, Billing and Project Cost Accounting. With some systems, you may also have the ability to add on additional components that integrate to form the entire ERP, including CRM, proposal management, project budgeting and resource scheduling, and possibly document management and business intelligence tools. Sheldon Needle, President and Founder of CTS, Rockville, Md., describes why a piecemeal approach to integrating systems can be a tricky undertaking:

Sometimes, companies attempt to create best-of-breed systems, selecting and integrating pieces of functionality to mirror the results of a single ERP package. Yet, unless you have in-house staff with the technological expertise to logically knit together disparate business systems, its better not to go down that path.

The quality of the integration is a very important factor in deciding if a particular solution will be easy to implement and train staff to use. It will also affect the quality of the reporting and extent to which spreadsheets will be required outside of the system. Some of the questions you should ask in the system evaluation process include:

- Did the vendor develop all of the key functionality, or did it acquire some of it and bolt it on?
- Are some of the modules or features supported by a third-party company?
- Are all the modules written in the same platform with the same technology and user interface?
- Can the integration be broken if the vendor does an update?
- Is the data integration two ways and automatic, or are you required to manually trigger the sharing of data?

The best systems are supported by a single vendor with a standard (non-proprietary) platform, backend database, and user interface. This will lessen the support required when there is a problem, accommodating the learning curve for training new staff, and any problems that can occur when updates come out. It may also minimize the hardware requirements and number of servers required if you are hosting the system yourself.

Evaluating the problems with your existing processes and systems, preparing a documented list of key requirements, and defining the KPIs that will help you achieve the ideal ROI will go a long way toward understanding whether a particular system or ERP will provide the benefits you want.

In tying the ROI evaluation of the software to the "Find the Lost Dollars" formulas, you can start to see that changing one or two of these metrics by even a small amount can have a substantial impact on firm profits. The key is to understand how the software will help you realize those changes in results and help you focus on them as part of the implementation plan.

I recommend that you employ the services of the software vendor or their partners to help you pinpoint the potential cost saving, increase in revenues, or improvement in efficiencies that you are trying to recognize. By understanding which of the areas your firm can gain the most benefit from the software, you can begin evaluating the potential ROI of the software. Whether it is win rate, recovery of extra services, or improved project management to reduce budget overruns and improve project profit margins—if you don't measure it, you won't see the desired results.

BEST PRACTICES FOR SELECTING NEW BUSINESS MANAGEMENT TECHNOLOGY

In order to get the desired ROI from new systems, you have to purchase the right system and experience a successful implementation. This can be a daunting task for most firm leaders, especially from a technology, and features and functionality standpoint. There are definitely some best practices for both selecting the software and implementing it successfully that I can share here.

Selecting the right software product for managing your business is obviously a lot more complex and involved than what I can elaborate on here. It could actually be a book of its own (and it is—I wrote it in 1990). A few key suggestions include:

Talk to your peers to find out what they are using and whether they love it – Vertical applications (ones that are specifically designed for A&E firms) are usually a much better fit. Many small firms use QuickBooks and stay on it much longer than they should. If you are using QuickBooks, then Microsoft Excel is really your accounting system. QuickBooks may actually hold your firm back from growing, and can cause issues with firm performance analysis and internal controls. For more perspective on when you should move off of QuickBooks, see my blog about how to know when you have Outgrown QuickBooks (http://acuitybusiness. com/blog/bid/176700/Five-Signs-You-Have-Outgrown-Quickbooks).

Check with one or two trade associations to which you belong – The AIA, SMPS, ACEC, and many others all have recommended systems and vendors.

Talk to your outside CPA firm – I greatly admire CPA firms and the services they provide to their clients. Most firms have an extremely high level of integrity and ethics and will always try to lead you to the solution that best fits your short- and long-term needs. One caveat though— your CPA needs to have expertise in both the A&E industry, and the products available that are specifically designed to work with the types of projects, billing, and reporting that you will need. Unless you are under 10 employees, QuickBooks is not a good solution, and will not meet your goals to grow.

I view the role of the outside CPA as a trusted advisor that can help with the implementation process, and ensure that your books are set up in a way to meet your tax and internal management reporting requirements. It is my opinion that if they are selling software for a profit, they are not independent, and run the risk of appearing to have a conflict of interests in making money from their recommendations. While it might seem to make sense to take advice from your auditor, his or her advice is only valuable if it is truly an independent auditor (not affiliated with a particular software vendor). That person also needs deep experience in helping A&E firms select and implement software.

Determine and document your requirements – Documenting your firm's requirements can be a time-consuming exercise, but it can also lead to a much more successful software selection. The first step in the process should be to document your current processes and interview your staff in all areas of the business to understand how they do their jobs every day. This is a similar exercise to the process documentation exercise in Chapter 5, but with a more detailed look at the specific automation requirements for each step. For example, when looking at time entry, you will want to focus on the detailed "features" that you may need to get the results you are looking for in your billing, reporting, and project management processes.

An example of how this should be done is to look at a critical function of the A&E business: time entry. Here is a list of just a few of the many timesheet management features you should consider:

- Capturing of time for time off, overtime, and special situations (marketing or promotional projects)
- Options for controlling what the employee sees, including access to projects/phases they are not budgeted on
- Approval processes (levels of hierarchy, options)
- Ability to record comments on each time record
- How the timesheet gets posted to the general ledger
- Controls and option for changing time records, both before and after billing and payroll
- Work breakdown structure and flexibility for recording time to specific codes for further reporting and analysis
- Warnings or alerts for employees charging time to the wrong projects or charging too much time
- Compliance with government contracts: audit trails, electronic signature, total time accounting (see Chapter 10)
- Real-time access to time records for project management and reporting
- Ease of making modifications, transfers and other adjustments for billing purposes
- Ease of remote access to timesheets outside the office and from multiple devices

This is just a sample of the many timesheet management features that you should consider when evaluating your requirements for a new system. You should not assume that all systems do the same thing; this is a dangerous trap that some software vendors try to cover up with the "smoke and mirrors" of the software demo.

No system out there is going to have 100% of the features you need "out of the box." If you can get 95% of your "must have" features, and

85% of your "want to have" features, you are doing well. You will almost always have a small amount of customization required, such as an invoice format or special report.

You must also take into account any special accounting, reporting, billing, or government contracting requirements (see Chapter 10) that will immediately rule out certain systems. Sometimes, the crucial features or reporting elements that you will need to satisfy your client's contractual requirements cannot be customized and must be 100% integral to the software system; otherwise, the long-term usefulness or scalability of the system is compromised. Some examples of this are multi-company or multi-currency capability, DCAA compliance, or ability to pay employees different pay rates for different work. If these types of requirements are deemed to be critical, systems that do not inherently manage these functions will not be a good fit for your business.

Ask for and talk with vendor software references – After shortlisting your potential software products to one or two, you should speak to at least three of their customers to understand their experiences in working with the software product, the software publisher, and the consultant or partner with whom you are considering working.

For best results, try to get firms that are around the same size, do similar types of work, and do not compete with your firm. Here is a list of some of the possible questions that you could ask a software reference:

- What products did you review during your selection process? Why did you pick this system?
- How long have you had it?
- How long did it take to implement?
- Did the implementation come in on budget and on schedule?
- Did you enjoy working with the consultants that you hired for your implementation? Would you hire them again?
- Describe any problems you experienced during the implementation

- Did your data get converted and did it meet your expectations?
- What would you do differently? Would you buy this product again?
- What do you like best and least about the software?
- What were your primary goals for the software purchase (metrics, efficiencies) and have you accomplished them?
- How have your employees embraced the change and do you have any recommendations for how to increase employee adoption?
- Were there any unexpected issues or costs that occurred during the implementation?

While these are just a few of the many questions that you may want to ask a software reference, you will get many insights into the real experience that these vendor's software customers have experienced when implementing the solutions you are considering.

An old story in the software industry illustrates the challenge of trying to assess a software product based on the demo. This is my interpretation of the story after many years of passing the story down:

A man dies and is suddenly confronted by an angel and a devil. He is offered two options: to spend eternity in heaven or hell. First the angel takes him on an elevator and proceeds up to a beautiful and peaceful oasis where he can spend his days relaxed and in perfect serenity. He is ready to choose this option, when the devil convinces him to have a look at hell, promising that it is not anything like what he has heard of before. He then goes with the devil down the elevator to hell. When the elevator doors open, he walks into a giant party, with beautiful women, music, food and drinks, and everybody is having a great time.

The man easily decides that hell would be a great place to live and tells the devil he has made the decision to stay in hell. As soon as he seals the deal, he is suddenly encompassed by burning hot flames and is forced to an after-life of hard labor. When the man questions the devil about why he is not able to live the party life that he first saw, he is told by the devil, "That was only the demo."

This story, while exaggerating the problem that exists with software sales processes in general, clearly puts the buyer in a "buyer beware" mode, and emphasizes the need for special care and due diligence when selecting a new software system.

SUCCESSFUL SOFTWARE IMPLEMENTATION

After the software has been selected, the really hard work starts. I often use the analogy of moving to a new house when describing a software implementation:

When you first start to move, you have a ton of work ahead. You have to pack all the boxes, mark them appropriately, and throw out what you don't need any more. There may be some cleaning and scrubbing you need to do to the old house. Then you must move all the boxes to the new house. You will usually need help with this.

Once you have all the furniture (data) and boxes (historical transactions) in the new house, you may feel uncomfortable. You may have to move things around a few times to get them where you want them, and you may not be able to find everything you need for a while. But after a while, you start to get things just the way you like them and start to get the benefits of having a nicer, newer and better house. In a matter of months, you have everything just the way you like it, and wonder how you ever survived in that old house for so many years!

While it's not possible to take away all the pain or reduce the learning curve that comes with moving to a new system, the following are some best practices that can effectively ensure that you have the basis for a successful implementation:

Executive commitment – In order to ensure a successful implementation of an ERP system, it is critical that the project has executive support and commitment. Without this, the accounting department (usually the "owner" of the implementation process) will be fighting an uphill battle in getting firm-wide buy-in and adoption. When everyone in the firms recognizes that the executives are stakeholders in the outcome of the system implementation, they are much more likely to embrace the new system and comply with new policies and procedures.

Treat the implementation as a project – Implementing a sophisticated ERP system, or even a basic accounting system for the first time is a lot of work. It can be complex for firms that are implementing many components and have a lot of entrenched processes they are trying to overhaul.

Just as you must plan and manage your billable projects to ensure they remain on budget and schedule, the software implementation must also be managed and controlled. Don't count exclusively on your outside consultants to manage everything for you. While they can provide guidance and suggested timelines, you must manage resources, problems, and activities internally to ensure that your company's goals are achieved.

Ensure you have realistic time expectations and resources – Do your homework and make sure you understand how long it will take to install, how much it will cost, and how much time it will take from your internal resources. With the right budget and schedule, your entire team will be more satisfied with the outcome.

One of the primary causes I have seen A&E firms not being able to achieve their desired go-live date is their inability to apply their resources to get critical milestones of the implementation process completed by the assigned deadlines. Just as with your A&E work, delays in the schedule can be difficult to make up, and almost always result in an extended timeline and increased costs.

So that will not happen, make sure that there are realistic expectations about the time commitment from your internal team members each week and that this time is allocated to getting the work done. If delays start to happen in the beginning, bring in additional resources to get the daily accounting work done so that your regular staff can work on the new system implementation.

Other "projects" should not get in the way and derail your staff from accomplishing their tasks each week. If people are getting pulled away for other activities, you may need to educate the entire team about the importance of the system implementation and the level of priority that it should take for the accounting staff. I very often see the firm executives pulling their accounting team off of the implementation in order to research something they need for a meeting. Commitment needs to be agreed upon and communicated across the organization about the potential costs and repercussions to the success of the project if this is allowed to happen.

One idea for bolstering your accounting team during a system implementation is to use college students as a resource for helping with a system implementation. They are smart, inexpensive, and usually more technology-savvy than normal temps. And depending on their majors, they may be very interested in the system implementation and can end up being great resources in the future as a summer intern or new hire when they graduate.

Set goals – Just like any other project, if you don't have well-defined and measurable objectives, then you won't have a way to measure success. By using the formulas in Chapter 2, and the suggested 6-step process in Chapter 11, you can identify where the largest gains will be and set some goals to find lost dollars in a few different areas of your company.

You will also need a timeline that will keep the entire project team on track. First you should determine your desired start date, and work backward from there. The implementation will take anywhere from 3 months for a small firm to 12 months for a very large system implementation for a firm that has more complexity and multiple locations. Each week of the implementation should be planned out in terms of what milestone will be accomplished and what each person's tasks will be. Completion of each step is dependent on the step before, so missing deadlines will have a critical impact on being able to accomplish your go-live date.

Hire the right consultants – A great consultant can make or break your implementation. Industry experience is very important. Some software publishers distribute their software through partners who may be more client-focused, local to your office, and better able to service your specific requirements.

There are basically two types of consultants: those who are "independent" and those who are biased toward one or more specific software products. My opinion is that no consultant is ever totally "independent." If they have been working with multiple software products for many years, they have already formed opinions about which products are better. Most consultants have their own agenda. It may be to stay involved for a long time during the implementation process and get "free training" as your firm pays the software company or product consultants for the implementation. In some cases they may even get a referral fee or "kickback" that they are not telling you about.

A "biased" consultant will be honest up-front about any relationship with a software company, and whether there's a preference for one product over another. As long as the consultant has the values and integrity to honestly discuss the shortfalls in his or her systems, and not push you toward a system that is not appropriate for your business, you will find the biased consultant may be better to rely on than a so-called independent one.

An important factor to consider in selecting the consultant with whom you want to work for your system implementation is to understand that person's implementation approach: Does he send someone to your office for a week at the beginning? Does she offer several hours a week over several months to work with you through the implementation? Will that consultant work with you on-site, virtually or in a combination of methods? Are you working with a team of subject matter experts or assigned to one person that is supposed to know everything about the system? Does that consulting firm have specific experience working with A&E firms, or does it work across multiple industries? Depending on your location, the modules of a system you are implementing, and your short- and long-term goals, the implementation partner will be a critical factor to your success.

Assemble an implementation committee – It is important to assemble a group of key stakeholders that will take ownership of the system implementation. This committee will determine the goals, manage the timeline for the project, and keep everyone accountable to getting their weekly "assignments" done on time. The committee should have at least one principal and a representative from each of the major divisions of the company: accounting, HR, project management, and marketing.

Develop new processes – Chapter 5 explained how important it is to have efficient processes that make your staff's jobs easier. If you continue to use the same old processes that you have always used (cultural trap 10) then you will get the same results. Don't try and make your new and more powerful system adapt to your old processes. Take the time to understand how to make the system more powerful and automate as many of the processes as possible.

Once you have determined your new desired processes, it is best to try and document them as completely as possible. If you are doing government projects, this may be a legally required aspect of your compliance with the Federal Acquisition Regulations (FAR). Documented processes help ensure that all company policies are thought through and available to employees for most common scenarios. It helps them understand approvals, policies for making changes to records, and "how-to" for most operational functions in the business.

Training, training, training – The best system in the world is useless if your employees do not know how to operate it. Even if it is implemented perfectly, it does not take long for it to get messed up by employees that are not operating it correctly. I have seen more systems fail for lack of training than any other reason. Role-based training is crucial for each person in the firm. You will not see your ROI from your system if you do not ensure that your teams are able to effectively use the system.

In addition, you should look at ways to do cross-training whenever possible. This is training multiple people on every aspect of the system so that turnover, vacations, or extended periods of leave do not adversely affect the operations of the company.

Try to record and document your training whenever possible. This helps in training remote staff or new staff if there is turnover, and ensures that everyone is getting the same training.

Change management – Changing a system can be stressful for some of your staff. Many of them do not like change and may see the new system as a burden. Explaining the reasons for the change and the goals the company has set for improving financial results is critical to success. You will see a much better attitude and approach by everyone if they fully understand why the firm is upgrading technology. Providing feedback about the progress of the implementation can also help keep everyone informed and hopefully allow them to see that the firm is making improvements that will benefit everyone.

Measure results – By setting goals in the beginning and measuring results at 6, 12, and 18 months after the system is live, you can determine if the expected ROI is being realized. In some cases, it may take longer to get the ROI. However, usually a year after your live date you will see some measurable results. I recommend that you start with your win rate, utilization rate, and project profit margin. Instructions for calculating and measuring the improvement to these metrics are provided in Chapter 11.

If you have many goals and areas that need improvement, you may need to take a phased approach to your implementation. By completing the online Find the Lost Dollars Assessment tool, you can start to analyze where the biggest gains and improvements can be realized from the system implementation and develop a plan for recognizing the attainment of these goals over time. For example, if you determine that the project profit margin and win rate are the areas where the biggest gains can be made, you may decide to focus on one at a time, by first working on project management and extra services automation, and then several months later implementing CRM and proposals automation.

While implementing a system can be a lot of work and stressful, the payoffs can be huge. Leveraging technology to make your people more effective at their jobs is the key to finding the lost dollars. Chapter 11 details a 6-step process to evaluating where and how to start finding lost dollars in your business.

Summary

- Many A&E firms are still using old technology to manage their marketing, operations, and financial management. Technology can help automate manual processes and make employees more efficient at performing their jobs.

- It is very difficult to keep up with all of the technological changes happening in the world today. Changes such as mobile devices, social media, and cloud computing present both challenges and opportunities to your firm, and should be evaluated to determine where you can find cost savings and increased efficiencies.

- Technology should be viewed as a strategic investment, and the ROI of technology purchases should be evaluated when considering where to make key purchases. Generally, leveraging technology is much more cost-effective than using people to fix problems and increase efficiencies.

- An ERP system is a system of integrated modules that allows data to be shared across the business to manage the entire project life cycle. Integrating marketing, accounting, and project management and planning into one system can allow the firm to grow faster and reduce the need for redundant data entry and spreadsheets.

- Selecting a software system to manage the business can be confusing and complex. It is important to analyze your specific requirements, and also talk with peers, financial advisors, and other industry resources to find the most popular applications.

- Successfully implementing a new system requires an organized and focused approach. It is important to have executive support throughout the implementation, as well as follow other best practices, including having a project plan, hiring experienced consultants, allocating and training internal resources appropriately, and setting goals and measuring success.

GOVERNMENT CONTRACTING

"Building Momentum in the government market depends on many factors, starting with your feet on the street—your people who meet with government clients and business partners."

~MARK AMTOWER, SELLING TO THE GOVERNMENT

I attended the *Zweig Letter* Hot Firm conference in Laguna Beach in 2011 with the primary purpose of finding the silver bullet to the success of the most successful A&E firms in the United States. The one thing that I came away with is that upwards of 90% of the firms represented did some kind of public work. These numbers certainly escalated during the recent recession, with stimulus money poured into infrastructure, and a war continuing overseas. Many A&E firms saw an opportunity to sustain and even grow their businesses from government projects.

But what many of these firms discovered is that over the last 30 years, through decades of government contracting expansion, forage into the government space is slow, competitive, and fraught with complexity. Government clients pose a whole different set of challenges for A&E firms, with the stimulus money now dry, and states making cuts because of declining revenues and increased deficits.

Government contracts are more competitive than ever, and with the long proposal cycles and firms cutting their rates, it is harder to make a profit on the jobs that you do win.

FINANCING GOVERNMENT CONTRACTS

Many startup businesses also find that they need financing to be able to cover payroll and other expenses for the first several months as they perform the work under the contract and have to wait to get paid by the government. The entire process from contract kickoff to the first cash receipt can be 60 to 90 days, usually requiring much more cash than the average entrepreneur can cover. George Nagy, Sr., consults with companies, helping them ensure they have the proper funding and infrastructure to bid and compete on government contracts. He warns new government contractors to have their financing in order before they win a government contract:

• •

Nine out of ten businesses are undercapitalized, of which 90% of the time leads to failure. For government contractors, getting a line of credit and or equity financing prior to contract award will allow you to afford the cost of "starting" up a contract. If you are new to government contracting, and do not have the expertise in properly billing to obtain your payments, you may not be able to float these cost which can be catastrophic in the early phases of your business. The costs which people may overlook are:

- *Employee Payroll*
- *Payroll Taxes*

- *Benefit Payments*
- *Telecommunications Cost*
- *Office Space*
- *Professional Services (Legal, Accounting, Marketing...)*

Having a line of credit in place prior to award will allow you to use your contract to obtain additional funds to ensure that you are ahead of the financial curve. If you fall behind, it is very hard to dig yourself out.

• •

As George advises, getting your financing in order before bidding and winning a contract will ensure that the company can afford the startup costs of growing the business and taking on the associated responsibilities in the short term.

THE COMPLEX WORLD OF GOVERNMENT CONTRACTING

In his popular book called *Selling to the Government*, Mark Amtower, Founder of Government Market, and guru of government contracting, observes the cruel facts about the limited success of firms trying to enter the government marketing:

• •

In my 28 years of studying the business-to-government (B2G) market, I have observed that about 90 percent or more of the companies that try to enter the government market fail. They fail not from lack of skill at what they do; they fail from a lack of understanding the nuances of a new market—a different market—with rules arcane enough to fill tens of thousands of pages of "government speak". They do not adapt." (Source: Selling to the Government: What It Takes to Compete and Win in the World's Largest Market, by Mark Amtower, 2011.)

• •

Mark's book is full of great advice, insider information and other valuable resources to help companies enter and succeed in the government space. Many firms have ventured into this space, competed, and won. However, as Mark points out, many more have spent incredible amounts of time and money and failed.

What you will find when you attempt to enter the federal contracting market is a myriad of rules, regulations, and processes that have to be followed in order to compete and win. The average new firm entering the government contracting marketplace takes a minimum of 18 months to get its first contract. The government has its own set of rules for almost everything a business does, which can present challenges in the following areas:

- Time and expense management
- Billing requirements, forms and processes
- Human resources and hiring rules
- Payroll and compensation limitations
- Benefits management
- Business insurance
- Recruiting and onboarding
- Resource management
- Marketing
- Proposal writing
- Pricing
- Contract management
- Cash flow management, banking and finance
- Financial reporting and analysis
- DCAA compliance (see below)

A whole world of experts have built their own businesses and livelihood around helping companies do business with the government.

Many of the traditional business management practices that you have employed in the commercial world will not work in government contracting, and having a good understanding of effective government contracting practices is essential to long-term success.

In order to illustrate this point, I have compiled a list of some of the challenges that our clients have to deal with in order to compete as a government contractor. This is not meant to be deep education about the government contracting process, just a high-level introduction. If you are interested in a very deep and thorough introduction, I recommend that you read Mark's book.

Registration – Government contractors must register to do business with the federal government. Most states also have a separate registration process to allow you to be a recognized vendor by the procurement officials in their state.

In 2012, the government began the process of combining eight federal procurement systems and the Catalog of Federal Domestic Assistance into one new system called the System for Award Management (SAM). As of November of 2012, it has completed Phase 1 of the project, which is called the entity management phase. Phase 1 includes integration of the Central Contractor Registration (CCR), Online Representations and Certifications Application (ORCA), and the Excluded Parties List System (EPLS).

Future Phases of the SAM include integration of the Federal Business Opportunities (FBO), Catalog of Federal Domestic Assistance (CFDA), Electronic Subcontracting Reporting System (eSRS), Federal Funding Accountability and Transparency Act (FFATA) Sub-award Reporting System (FSRS), Wage Determinations Online (WDOL), Federal Procurement Data System – Next Generation (FPDS-NG), Past Performance and the Information Retrieval System (PPIRS). If it isn't included in this list, it is not planned to be part of SAM.

Users that had previously registered in the CCR will now have to get a new user account at www.SAM.gov. All information previously entered in the old CCR system has been transferred to SAM. The remaining phases will be done over the next few years.

Certifications – The government gives preferences to small and/or disadvantaged businesses in order to "level the playing field" and give these groups a fair chance to compete for federal money. The government will set aside certain contracts, types of contracts, or percentages of contracts for companies that meet specific criteria such as Minority Owned Small Business (also related to the specific Small Business Administration (SBA) program called the 8(a) program), Service Disabled Veteran Owned Small Business (SDVOSB), Woman Owned Small Business (WOSB), or Historically Underutilized Business Zone (HUBzone). Some of the certifications last forever while some, like the SBA 8(a) program last for a certain number of years, and have specific growth criteria attached to them.

Some of the certifications must be officially confirmed with the government, and others are self-certified. The rules, terms, and conditions for these certifications can change at the preference of the government. such as what amount of revenue or employees determines a small business. Keep in mind that the criteria and certification process may be different for each state and industry if you are pursuing state government contracts.

Federal Acquisition Regulations (FAR) – these are the official legal rules that government agencies must follow in procuring goods and services, and the specific regulations that contractors must obey when bidding and performing under a government contract. The FAR are an extremely complex set of regulations, taking up volumes in written form. For this reason it is recommended that you work with a CPA or other professional that specializes in understanding and applying the FAR to your company's accounting practices, and management of your government contracts. Failure to comply with the FAR can result in severe financial setbacks.

Contract Audit Standards (CAS) – The CAS is another set of regulations that seek to provide consistency in the way that government contractors manage their books and bill the government. They seek to apply specific criteria and principles to the measurement of costs, the matching of costs to the correct accounting period, and the allocation of costs to the cost objectives (general ledger accounts or projects).

Contracting Office (CO) – This is the buyer for the government agency. They must follow the regulations as set out in the FAR and CAS in order to legally purchase goods and services from vendors and contractors. They can determine if and when a contractor should be audited to ensure compliance with the FAR.

Defense Contract Audit Agency (DCAA) – Because of the complexity of the FAR and CAS rules, the government must audit contractors to ensure compliance and accuracy in contractors billing to the government. When an audit is mandated by the CO, the DCAA will be requested to complete a system audit, pre-award audit, or annual incurred cost audit. While the DCAA was originally created to audit only U.S. Department of Defense contracts, they now get called into audit contracts for other agencies.

Failure to pass a DCAA audit can result in harsh financial penalties, including refusal to pay certain contractor claims (invoices), loss of contracts all together, or in the case of intentional fraud, jail time! For this reason, compliance with the FAR and DCAA accounting practices is critical in managing federal government contracts.

Contract Vehicles – The government has many ways that it can procure goods and services from government contractors. These range from buying products or services with a credit card, to GSA schedules and Indefinite Delivery, Indefinite Quantity (IDIQ), to awarding large cost plus contracts that are complex and highly scrutinized. The government must have a contract vehicle in order to buy goods and services, and there are many different ways that they can initiate a transaction.

Getting educated on the different types of contract vehicles and teaming with companies that already have access to these vehicles is a good strategy for successfully navigating the complicated GovCon waters.

Cost Plus Contracts – While most A&E firms are familiar with T&M and fixed fee or lump sum contracts in the commercial world, most will never have the opportunity to bid or win a cost plus contract unless they pursue a federal government contract. Consider yourself lucky if you never get one of these!

In its simplest terms, a cost plus contract allows the contractor to bill for all costs that are incurred as a result of doing business. This includes all direct costs—the labor and ODCs directly attributable to the work on the contract, and the indirect costs—the fringe benefits, overhead and general and administrative labor and expenses required to manage the business.

Because contractors are charging the government and getting reimbursed for all of their incurred costs, the government must audit the contractor annually to ensure that the contractor is following the FAR and CAS, calculating all of their costs correctly, and not billing the government for any unallowable costs (see FAR 31.201).

Winning a cost plus contract ensures that your accounting practices, internal control processes, and level of scrutiny over your books is going to change dramatically! For this reason, I recommend that contractors who have never had a cost plus contract give serious consideration as to whether their internal processes, systems, accounting staff, and infrastructure are ready to take on this level of compliance obligation. Most of our clients who have cost plus contracts must have a CFO or trained accounting professional on their staff, as well as a contract management professional. While these roles can be outsourced, it is highly recommended that resources with this expertise be part of your pricing, bidding, and contract management team.

Segregating Direct and Indirect Costs – A major accounting principal in government contracting is proper handling of costs. In the case of a cost plus contract, costs must be segregated by the accounting system into direct and indirect cost pools. This includes the detailed labor charges on the timesheet being divided up by billable project, and into all of the indirect pools such as holiday and paid time off (PTO), bid and proposal (B&P), administrative time and other non-billable activities.

While this can be done using QuickBooks, in conjunction with quite a few spreadsheets and manual processes, you should consider getting a system with effective cost accounting capabilities that can withstand the scrutiny of a DCAA audit. In addition to the project accounting requirements and labor distribution requirements, there are many other detailed time entry requirements, approvals, and other processes that must be followed in order to be compliant.

Once the costs have been segregated, the indirect rates of the company must be calculated and submitted to the government in the competitive bidding process. If mistakes are made in the rate calculations, or the assumptions that were used to calculate the rates were wrong, the government can come back and require the contractor to pay back money that was incorrectly billed to the government.

Mark Amtower has also offered this insightful anecdote that demonstrates the degree to which your company's business risk escalates after winning a government contract:

* *

Over the years I have had hundreds of "emergency" calls, most indicating that a contractor or would-be contractor was in need of the right outside help. On one such call, a woman was referred to me by a mutual friend. She called in a big panic. After more than two years of pursuing government business she had finally won a small contract, but the government was not paying for her services.

During a brief chat, I figured out that she was sending a standard invoice to her client, totally out of compliance with her contract. I referred her to her local Procurement Technical Assistance Center (PTAC), which helped her through the billing process.

Education is the key to this market, especially for "newbies". Company infrastructure must be ready for GSA and DCAA audits and more. The more educated a company executive is, the less likely it is that they'll be making "911" calls. The moral of this and of most of my "911" calls: whether it is bid and proposal, accounting, legal, or marketing, get advice from the right professional(s) before proceeding in the government market.

• •

As you can see, doing business with the federal government can be expensive, time-consuming, and even risky! You had better know what you are doing before venturing into this unknown territory.

Guy Timberlake, Founder and Chief Visionary of the American Small Business Coalition (ASBC), has devoted his life to helping small businesses do business with the federal government. He is an expert on all aspects of small business government procurement and frequently writes in his popular government contracting blog about the advantages of small businesses taking advantage of the Simplified Acquisition Threshold (SAT) for government procurement. The SAT is a contract vehicle that attempts to simplify the process for a small business to get up to $150,000 in federal contract money without the usual complex competitive bidding process.

His November 18, 2012 blog post entitled, *Instead of gazing out over the 'cliff' let's turn our attention to $500M in simplified spending,* urges small businesses to look at the SAT as a way to get federal money in the shadow of the looming sequestration cuts. (Source: http://www.theasbc. org/news/109031/Instead-of-gazing-out-over-the-cliff-lets-turn-our-attention-to-500M-in-simplified-spending.-.htm)

In the article, Guy expresses his dismay and concern at the focus of small businesses toward large contracts that are difficult to win, rather than the SAT money that is more plentiful, faster, and easier to get:

• •

> *It seems folks are still aiming at the large multi-year and multi-million (and billion) dollar opportunities and hunting licenses, versus capitalizing on an area of spending that has increased dramatically since FY2003. After all, these procedures were created in part to "Improve opportunities for small, small disadvantaged, women-owned, veteran-owned, HUB Zone, and service-disabled veteran-owned small business concerns to obtain a fair proportion of Government contracts."*

• •

Guy's blog post also references a letter from the Executive Office of the President, Office of Management and Budget (OMB), that specifically recognizes the fact that small businesses are not using the SAT program to the extent that the government expected or desires, and talks about how these opportunities are being lost to larger companies. The following is an excerpt from this letter:

• •

Maximizing Opportunities for Small Businesses Under the Simplified Acquisition Threshold

Pursuant to longstanding statutory requirements in the Small Business Act, agencies are required to automatically set aside work for small businesses that is equal to or less than the value of the simplified acquisition threshold (SAT) (generally $150,000) unless the contracting officer determines the "rule of two" cannot be met – i.e., there is not a reasonable expectation of obtaining offers from two or more responsible small business concerns that are competitive in terms of market prices, quality, and delivery. However, a third-party analysis of data in *the Federal Procurement Data System suggests that a significant amount of work under the SAT is not going to small businesses, including for products and services in industries where small businesses are typically well represented. This suggests that opportunities for small businesses are being lost, and that agencies must take additional steps to consistently apply set-asides in the manner prescribed in law and regulation. (Source: http://www.whitehouse.gov/sites/default/files/omb/procurement/memo/follow-up-april_25-2012-meeting-of-the-small-business-procurement.pdf)*

• •

This example clearly illustrates the advantages of doing your homework, understanding the complex government contracting landscape, and surrounding yourself with peers and advisors who can help guide you towards success.

GOVERNMENT CONTRACTING RESOURCES

Anyone who is already working on federal contracts or considering pursuing federal contracts, and is also a small business, should join the ASBC and get some face-to-face time with Guy Timberlake. In addition to helping with strategy, business development, proposals and contract compliance issues, the ASBC is a great networking group for companies looking to get educated and potentially team with other small businesses (go to www.theASBC.org for more information). This will be time well spent and you will walk away with solid ideas and advice that you can use immediately.

In addition to the ASBC, I recommend the following resources for advice on working in the federal government sector:

Selling to the Government: What it Takes to Compete and Win in the World's Largest Market, Mark Amtower, John Wiley & Sons Inc., 2011.

Government Contracting for Dummies, Deltek Special Edition, John Wiley & Sons, Inc., 2012

System for Award Management (SAM) – https://www.sam.gov

Official Government Website – http://search.usa.gov/

GovWin – A Web site dedicated to helping government contractors find federal, state, and local opportunities – www.GovWin.com

The official blog of Guy Timberlake and the ASBC – http://www.chiefvisionaryblog.com/

Federal Acquisition Regulations (FAR) – https://www.acquisition.gov/far

Federal Register – www.gpo.gov/fdsys

Business Partner Network – www.bpn.gov

Federal Procurement Data System – https://www.fpds.gov

Defense Contract Audit Agency (DCAA), an excellent resource for learning more about the audit process www.DCAA.mil

Acquisition Central – https://acquisition.gov

Federal Agency Procurement Forecasts – https://www.acquisition.gov/comp/procurement_forecasts/index.html

FedBizOpps – www.fbo.gov

Federal Acquisition Jumpstation – http://prod.nais.nasa.gov/pub/fedproc/home.html

Society for American Military Engineers – www.same.org

Procurement Round Table – http://www.procurementroundtable.org/index.html

PRO-*Net* – Small business database accessed thru small business search – http://dsbs.sba.gov/dsbs/search/dsp_dsbs.cfm

SUB-Net – Subcontracting opportunities posted by large firms – http://web.sba.gov/subnet

Office of Small and Disadvantaged Business Utilization (OSDBU) – www.osdbu.gov

FedSpending – Provides data on federal contracts by contractor, location, and agency – www.FedSpending.org

USA Spending – www.usaspending.gov

Government Finance Officers Association – www.gfoa.org

Professional Services Council – http://www.pscouncil.org/

Military Installations – www.armedforces.com

Federal Executive Boards – www.feb.gov

National Contract Management Association (NCMA) – www.ncmahq.org

GovEvents listing – www.Govevents.com

The Excellence in Government Conference – www.excelgov.com

Government Executive – www.govexec.com

Federal Gateway – www.fedgate.org

FedWorld – www.fedworld.gov

Where in Federal Contracting – www.wifcon.com

Federal Design & Construction Outlook Conference – www.aececobuild.com/the-federal-design-construction-outlookconference

DBIA Federal Sector Conference – www.designbuildfederal.com

Summary

- As a result of the economic downturn over the last six years and the increased spending by the federal and state governments on infrastructure, more A&E firms have been pursuing government contracts.

- When pursuing a large contract, it is important to secure adequate financing in order to be able to cover the cash requirements of the first several months.

- There is a great deal of complexity involved in bidding and working on federal government contracts. Many rules and regulations have to be followed, as well as terminology and acronyms to be learned. In addition to the myriad of Web sites and places where you have to register to work with the government, there may be requirements about your financial management processes and how you manage your accounting systems.

- If you bid on a cost plus contract, DCAA will probably audit your firm. Compliance with FAR is critical to staying in compliance with the DCAA. Accounting for a cost plus contract requires that direct (project related) and indirect (overhead) costs are segregated into the appropriate cost pools and projects according to the FAR.

- There can be significant time required and a steep learning curve to win your first federal government contract. You should consider hiring experts to help with strategy, pricing, bidding, proposal generation, and compliance to ensure success navigating the complexity of government contracting.

- This chapter provides some resources, including Web sites and organizations that can help you be more successful in government contracting. It is imperative to fully research and understand the commitment that you are undertaking when pursuing federal work for the first time.

HOW TO FIND THE LOST DOLLARS IN YOUR BUSINESS

6 Steps to Find the Lost Dollars in Your Business

"Your success in life isn't based on your ability to simply change. It is based on your ability to change faster than your competition, customers and business."

~Mark Sanborn, Author and Leadership Consultant

We have covered a lot of ground in this book, and hopefully, you're now inspired to take a different look at your business and encouraged to do something with this new perspective. If that is the case, I have done half my job. The other half is helping you understand how to take all of this information and apply it to find, and recover, the lost dollars in your business.

To recap what we have covered so far, we have looked at the cultural traps in the typical A&E firm and how they affect your profitability. We then took a detailed look at nine areas of the business where money is wasted or revenues are not maximized and the potential of improving them. We then reviewed the different areas of the business and many best practices that I and numerous other noted experts have developed over the last 24 years.

The final step in this process is to apply all of this knowledge to your own business in order to make a measurable impact on your bottom line. Some firms will see fast results. This is because certain changes can have an immediate impact on profitability. A good example of this is with time management. If your current time management processes are flawed, making the changes recommended in Chapter 6 will deliver a quick impact on many areas of your business. Other changes such as revamping your project budgeting practices require a larger investment, including training, development of new processes, and holding people accountable.

In any case, change is difficult. It requires conviction, commitment, and hard work. It also requires excellent communication, and careful planning. Using the analogy of moving to a new house that I used in Chapter 9, you can see that while the effort is painful, the gain at the end is well worth the effort.

My personal goal in writing this book is to help you find the lost dollars in your business. To make this task achievable, I have laid out a 6-step process that can be used over and over again to improve and eventually fine-tune your business management effectiveness. I have narrowed the focus of this section down to three primary metrics: win rate, utilization rate, and project profit margin.

While we discussed many other measures of operational effectiveness such as AR turnover and the time lost from inefficient processes, I believe that by focusing on the changes needed to improve these three primary metrics, many of the other problems in the company will improve accordingly. An example of this concept is that if you improve your project management practices to positively affect your project profit margin, you will naturally bill more extra services. All of the steps involved in improving your numbers are achievable and have a great payoff.

You are not alone in this endeavor. I am here to help you implement this 6-step process! In addition to the detailed instructions provided below on how to execute on these six steps, I have developed some online tools that you can use today to start finding the lost dollars in your business. These tools will facilitate the calculations that you need to do in order to understand your own situation better, determine where to focus your efforts, and measure your success as you put my recommendations in place.

The 6-step process that I have developed to help you find the lost dollars in your business are listed here and described in detail below:

1. Calculate your metrics
2. Identify where the most money can be found
3. Analyze the components that affect that metric
4. Develop a plan to make improvements
5. Implement changes
6. Measure and track your results

STEP 1: CALCULATE YOUR METRICS

In order to implement any project, you need to define your starting point. Look at this component of this project as the site survey. You are going in to figure out what you are working with. In some areas of your business you will find good things that are working. In others areas, you will find room for improvement. The purpose of this exercise is to do the site survey of your business and document its present condition.

This is just your starting point. If you find numbers that are not satisfying, then approach this as if they can only get better! This step will require you to do a little research. For firms that are automated with project accounting capability, this part will be easier. For those of you with QuickBooks and spreadsheets, it may be a bit more work. But this work is critical to uncover where the money is hiding in your business, and start the process of capturing it.

Your job in Step 1 is to calculate three metrics: your win rate, utilization rate, and project profit margin. All three of these metrics are expressed as a percentage. I have provided both the data points that you need to find, as well as the formula to calculate your metrics. In order to do this, you will have to dig into your financial data and gather the following figures:

Win Rate

- Number of proposals submitted
- Number of proposals won

> **FORMULA:**
> Win rate = number of proposals won ÷ number of proposals submitted

Example: Win Rate = 30 proposals per year won ÷ 100 proposals submitted

Utilization Rate

- Total direct hours for all billable staff
- Number of billable Employees

> **FORMULA:**
> First calculate the total standard hours for your billable employees, and then calculate the utilization rate:
>
> Total standard hours = Number of Employees x 2,080 hours
>
> Utilization rate = total direct hours ÷ total standard hours*
>
> *You can modify this calculation several ways. One common way is to subtract PTO and holiday hours from standard hours. Consistency is the most important thing when looking at this metric. For employees that did not work the entire year you must calculate the standard hours for the period of time that they were employed.

Example: Total standard hours = 70 billable employees x 2,080 = 145,600

Utilization rate = 87,360 ÷ 145,600 = 60%

Project Profit Margin

- Average project revenue
- Average project profit

> **FORMULA:**
> Project profit margin = average project profit ÷ average project revenue

Example: $25,000 ÷ $250,000 = 10% profit margin

STEP 2: IDENTIFY WHERE THE MOST MONEY CAN BE FOUND

Based on the results of your metric calculations, you should have some good numbers to start analyzing where money is being lost in your business. The next step involves calculating a 1% improvement in the metric, and determining which area will give you the biggest ROI to address. This is done by looking at the total dollars being lost based on your current volume of proposals or projects being executed on. Once you have done this analysis, you can better understand where the most money is hiding and start putting a plan in place to find it.

In order to do this, you will need to calculate the following figures:

- Average dollar value of proposals submitted
- Number of billable employees
- Average billing rate
- Average project revenue per year
- Number of projects per year

Now plug these numbers into the following calculations to determine the impact of a 1% increase in each metric:

Win Rate

> (Number of proposals submitted x .01) x average proposal value = impact of 1% increase in win rate

Example: 100 Proposals Submitted x .01 x $370,000 average proposal value = $370,000 per year

Utilization Rate

> (Total standard hours x .01) x average billing rate = 1% increase in utilization rate

Example: 145,600 x 0.01 x $110 = $160,160 per year

Project Profit Margin

> (Average project revenue per year x number of projects worked on) x .01 = 1% increase in project profit margin

Example: ($250,000 x 40) x .01 = $100,000 per year

Depending on the size of your firm, the number of proposals submitted and the number of projects worked on each year, you may have some significant numbers. Most firm owners are shocked to see these numbers and realize the money that can be gained by improving business management practices and automation.

Chapter 2 detailed nine areas where money gets lost; however, in this chapter we are only looking at three metrics. This is because some of the nine areas will contribute to one or more of these three metrics. An example of this is how lost opportunities and the cost of poor proposal development processes both affect the win rate. So in several cases, more than one of these nine areas will contribute to the results of one metric. A few of the nine areas do not contribute directly to these three metrics. However improving them will also impact the bottom line in a positive way but may be more difficult to measure.

The final aspect of Step 2 is to set some goals. In this exercise, you calculated the impact of a 1% change in each these three metrics. However, that may be too low of a goal for you. Maybe you should be attempting to realize a 2%, 3%, or 5% increase. Based on your knowledge of your business, you should have a better idea than I do as to what a reasonable goal should be for you to pursue. I will leave this up to you, but I ask you to document what your exact goals are for improving each of these metrics, so that you can measure and track your progress and improvement in Step 6.

STEP 3: ANALYZE THE COMPONENTS THAT AFFECT EACH METRIC

Now that we have a baseline to start from, we can begin to put some perspective on the value of this exercise. If the numbers that you calculated do not represent a significant amount of money to you, or are close to where you believe you should be, then congratulations! You are one of the few firms that cannot see a benefit from improved processes and systems.

However, if you are not happy with these results, and discovered that a 1% improvement in any of these metrics could mean some real revenue increases or cost savings, then hopefully you are prepared to continue with the remaining steps to find and recover the lost dollars in your business.

Step 3 is probably the hardest. This is where you take a deep look at your business management practices, and understand where the money is getting lost. You may have to examine several areas and conduct some analysis of the numbers. You may have found in Steps 1 and 2 that determining the numbers was difficult to do. If that is the case, then this should emphasize to you that you are not receiving the information that you really need to run your business.

By going back to Chapter 2 and the nine areas where money typically gets lost in A&E firms, we can start to develop an understanding of how we can improve these three metrics, as well as some other changes that will help increase the bottom line. While some of these nine areas do not have a metric directly tied to them, such as non-integrated databases and poor management of client relationships, it is my belief that the changes you will make to improve the nine areas will have a direct positive increase in all three metrics calculated above.

In this step, we will take a deep look at your current business practices that relate to the nine areas reviewed in Chapter 2. This will help us analyze how well you are managing or not managing each of them. I recommend that you look at each of these areas as a source of lost money, and try to determine the extent that processes are not efficient or systems are not integrated and meeting the needs of your team. I have provided some guidelines for understanding each area better, as well as questions to ask yourself and your team about how well you are managing each area.

Remember that at this stage of this evaluation process, we are only collecting the information that we need to move on to the next step and are not trying to fix everything that is wrong as we find problems. In order to effectively make a measurable difference in improving our metrics, we need to have the whole picture, look at how the various issues tie together, and create a plan for fixing them.

I also urge you not to get too panicked or upset as you go through this process. You will find some things happening that you did not expect or know about. This is normal and shows that you are making progress toward making the improvements that need to be made.

In order to collect and record the data that you will uncover in this step, I urge you to start creating a document that will become the basis for understanding what needs to be done to improve your business operations. I have created some templates that you can download from www.FindtheLostDollars.com that will enable you to collect the data and use later to develop your plan in Step 4.

Cost of Lost Opportunities

In order to analyze this function of your company, you need to look at how opportunities get into your system (if you have one), and how they are taken through a sales process to a close. I suggest the following process for analyzing your opportunity management process:

Interview each team member who has any role in the marketing, business development, or sales process. In these interviews, you want to ask the following questions:

How are leads about new projects found and where do they come from?
Do we have a Go/No-Go process? How is it done, and who is responsible for using it? Get a copy of what is being used and determine if it is effective. Interview your employees to get your feedback about its usefulness and value.
How is data about opportunities being tracked and managed?
Is there a written process for managing opportunity data?
Who is responsible for entering, managing, and updating the data about open opportunities?
If you are using a system, how often is the data updated?
What systems are we using to track opportunities?
Do we have a documented sales process? If so, what shortfalls are occurring in it?
Who has access to change or delete the opportunity data?
Can we determine our projected pipeline at any time?
Are we using probabilities to estimate what we will win?
Can we analyze which types of projects we have better success with?
What bottlenecks occur in the process that take time or lead to ineffective follow up on opportunity deadlines?

What reports are being generated to analyze our results, and what metrics are being analyzed each month?
Are we tracking win-loss data such as competitor information? Who we are losing to and why?
How are deadlines and stages of the sales process tracked?
Whose responsibility is it to ensure that all deadlines are met, and all follow ups are done?

In addition to the interviews, the quality of the data should be analyzed to understand how accurate it is, and whether enough information about each opportunity is being collected. If your win rate is very poor, further analysis should be done to uncover what could be done better. Employees on the front lines of this process will usually have the best feedback and ideas.

Lost Revenue from a Cumbersome Proposal Process

As described in Chapter 2, the win rate is also impacted by the proposal preparation process. If we are pursuing the right projects, the other factors that go into preparing winning proposals include how well they are targeted, the relevance of the data provided about our past performance, how professional and aesthetically pleasing they look, and even how many proposals a month or year we are preparing. Both the quality and the quantity of our proposals are affected by how difficult it is for our proposal teams to get the information they need, and how cumbersome it is to put it all together in a format that will impress the prospective client and beat the competition.

In order to understand how cumbersome your current proposal processes are, you must interview your proposal team, including anyone in your firm who works on proposals. The following questions can give you a great deal of insight as to how effective your current processes and systems are:

Do we have a database of employee qualifications and project data that our proposal teams can search and access for proposals?
Is any of the data in word documents, spreadsheets, or in people's heads?
How efficient is the sharing of data between the accounting staff and the marketing staff?
Who works on proposals and what training have they had?
Is there consistency between teams in how proposals are formatted and written?
How long does it take to prepare an average proposal?
How many proposals are you submitting per month?
How cumbersome is the process for assembling the data in the proposals?
Do they have to reinvent the wheel every time they do a new proposal?
Is there a lot of cutting and pasting to prepare a proposal?
When someone leaves the firm, do they take a lot of company knowledge with them?
Can we easily access photos of projects? Are they well organized?
Are we tracking data about subcontractors, competitors, clients and other team members? Where is this data managed?
Is it easy to locate the appropriate boilerplate documents for use in a proposal?
Do we have templates for resumes, project detail sheets, etc. that are easy to use?
Is there consistency in the proposal formats between teams?
Is there consistency in the naming conventions used for proposal file names?
Are proposals easy to find on the server?
Do we customize each proposal or use a few standard templates for every proposal?

Is there collaboration and sharing of data between multiple teams or offices?
Have we submitted two bids on a project from two different offices?
Is there a formal approval process before proposals are submitted?

I also recommend that you sit in with your proposal team from start to finish as they work on a typical proposal. Understand what frustrates them, what is taking too long, and the difficulty in the actual development of the proposal from beginning to end.

Lost Revenue Due to a Flawed Estimating Process

As we discussed in Chapter 2, the success of our project very often starts with the estimate. This is where we define the work to be completed (the scope) and the cost associated with completing each of the detailed milestones of the project. Without a good estimating process, the project is often destined for failure before we win it. In order to evaluate the effectiveness of your estimating process, you need to get the answers to the following questions:

Do we use standard templates that are managed and updated by the accounting department or contract managers?
Is there consistency between estimators in the language used to describe each phase, task and activity?
Who is involved in estimating?
Do we have written policies for creating estimates?
Are hours being estimated for each phase and task?
Are the templates being used updated regularly to ensure they have the most recent billing and/or cost rates?
Are we adding line items for meetings on each phase?
Are we calculating the estimated hours and labor categories that it will take to accomplish each phase and task, and then applying budgeted hourly rates to each hour by labor category to calculate the cost of labor?

How are we estimating the direct expenses required including travel, blueprints, delivery, etc.?
How are we estimating the cost of subcontractors on each phase and task?
Are we using the correct provisional overhead rate to burden the direct costs?
Are we adding in a margin for contingencies?
Do we use a consistent process for creating estimates or is everyone doing it any way that they want?
Is there an estimating system or are they being prepared in spreadsheets?
Are we using historical data or just cutting and pasting from old estimates?
Are estimates being reviewed or approved by anyone other than the preparer before being submitted?
Is the estimating data being used to form the budget if we win the project?

By getting the answers to our questions, we can start to take a deeper look at where our projects are not being estimated as accurately as they should, and why we may be experiencing budget overruns down the road. In addition to these questions, I recommend that you do an evaluation of the actual templates being used to prepare estimates. If templates are not being used, then an evaluation should be done of the process (or lack of) that is being utilized to create your bids.

Lost Revenue from Extra Services That Are Not Recovered

One of the big reasons for projects running over budget is the failure to bill for extra services. Extra services are commonly caused by what is called scope creep, which is the practice of performing services outside the scope of the contract without a contractual modification.

There are several poor project management practices that cause scope creep, eroding our project profitability. In order to understand the causes of scope creep, you must evaluate several major areas of your project management processes:

- Communication about the scope with the team
- Written policies for dealing with out-of-scope requests by clients
- Detailed communication with clients according to the company policies
- Time management and approval processes
- Setup of projects to capture extra services

As we learned in Chapter 2, up to 60% of A&E firms fail to communicate the complete scope of services with their teams. If your employees do not know what they are supposed to be doing, how do you expect them to do it? The following questions are designed to help you evaluate your current handling of extra services to determine if the processes and systems you are currently using are enabling you to reduce scope creep and/or bill for all of your extra services. I advise you to not assume that you know the answers to these questions. Take the time to interview staff members from each group in your office. The more people you talk with, the better insight you will have as to where the problems lie:

Do you have adequate discussions and documentation with clients about their expectations for extra services requests?
Do you have well-documented policies and processes for managing and controlling extra services requests?
Does your contract include language that matches your company's policies?
Are these policies communicated to your clients? How?
Are these policies communicated to your staff?

Does your staff understand and follow these policies? Ask for examples of how they have handled these situations in the past.
Do your policies include an automated or paperwork process for approval by both the client and the PM for extra services before the work is performed?
Do you always have a kickoff meeting with the team to review the project?
Who is included in the kickoff meeting?
Does the kickoff meeting include review of the contract including the scope, schedule, budget and deliverables?
Do you inactivate phases and tasks as soon as they are completed?
Do you inactivate projects as soon as they are completed?
Do you use special codes or projects to identify extra services on the timesheet?
Do your employees know how to charge extra services on their timesheet?
Do your employees know how to communicate with clients about your internal handling of extra services requests?
Are the policies for tracking and billing for extra services being followed?

In addition to interviewing your staff, I recommend that you look at your contracts, sit in some of the kickoff meetings, and review your company's timesheet practices for charging of extra services to projects. This is an area where many firms lose money. It is worth your time to understand the real life problems that are causing this money to be lost.

Cost of Low Utilization and Poor Resource Scheduling

There is a balancing act occurring every day in your firm between ensuring your employees are utilized effectively and making money on projects. Some firms have many small projects, and others have fewer large projects. This will affect how easy or difficult it is to predict and use your resources according to a plan.

Yet the payoff from managing this aspect of your business can be huge. Just a 1% increase in your firm's utilization rate is usually a significant number, and this surprises many firm owners when they do the calculation. If you believe that a 1% increase in your utilization is attainable, and the number is significant, then this may be where you want to start in the process of finding lost dollars in your business.

As we discussed earlier, one of the benefits to attaining optimal utilization of staff is being able to look into the future and analyze requirements for potential hires or reduction in staff. By being able to accurately predict when you will need to hire or reduce your staff, you will have the best chance of keeping your team optimally utilized and make the most money. The following questions can give you some insight as to how well your firm is managing resources and forecasting future requirements:

Are projects planned into the future with estimated hours for each person or labor category by time period (week or month)?
Do you have spreadsheets or an automated resource scheduling system that ties in with your budgeting process?
Are projects that are expected to be won factored into scheduling?
Are plans updated weekly according to changes in assignments based on project progress, delays and changes to the scope?
Are the schedules reviewed at Monday morning meetings?
Does each employee have a target utilization rate, and are they provided the data showing how well they are achieving it?
Do you have a firm-wide utilization target, and an individual target for each department or discipline? Are you reporting on and analyzing utilization on a frequent basis?
How do you balance the assignments for each staff member (over- or underscheduled)?
What scheduling or forecast data is being used to make hiring decisions?
How quickly are decisions made about reducing the staff due to lack of work?

| What other adjustments are made when workload forecasts are not adequate to keep existing employees billable? |
| Are timesheets approved frequently (weekly or bi-weekly) to ensure that employees are working on the correct projects/phases and not exceeding estimated man hours? |

A good look at your resource management practices can have big payoffs. While this may be very difficult for you to manage based on the types of projects you have, finding a way to control your employees hours on projects according to the estimate, and gaining visibility into future labor requirements will go a long way to helping you optimize your labor resources, and increasing your utilization rate.

Cost of Poor Project Management

As I explained in Chapter 2, there are three primary reasons that projects lose money. They are:

- Poor project budgets
- Failure to use resources as planned
- Unexpected changes in the project scope, resources or outside factors

Because utilization and project profitability are often directly opposed to each other, the management of projects must be carefully orchestrated to ensure that everyone is working toward the best result for the client and your firm. In looking at your firm's practices around budget development, management of resources against the budget, and how well your budget allows for unexpected changes, you can start to understand the issues faced by your managers.

In Chapter 5, we looked at the tough job of the PM, and I gave you some recommendations for improving their performance through better tools, training and assistance. The first step in evaluating your company's internal project management practices is with your PMs, and the reality of a day in the life of a PM at your firm. Here are some questions, interviews, and assessments you can do to understand how productive their processes and systems are in controlling project costs and ensuring projects are profitable:

Does the PMs job description accurately describe all of their responsibilities (see Chapter 5 for a list of typical PM duties)?
Have they had training on all aspects of their responsibilities?
Is there a formal budgeting process in place for developing, updating and monitoring budgets on a weekly basis?
Are you looking at project performance by PM?
Do you have the right people in the PM position (based on performance analysis)?
Is everyone using the same process for budgeting?
Are budgets ever done "on the fly" based on the amounts previously estimated on other projects?
Do your budgets include hours by person or labor category for each phase and task on the project?
Do your PMs approve all extra services before the work is done?
Do you ever start working on projects before you have a contract?
Do you have detailed, documented timesheet procedures that every employee is required to follow?
Are your employees filling out their timesheets on a daily basis?
Do your PMs rigorously approve timesheets?
Are your PMs frequently using more expensive people than they had budgeted?
Are the PMs doing too much of the work themselves?
Do your PMs have annual, quarterly and monthly revenue and profitability goals and metrics that they are being evaluated on?

Does your firm have an automated budgeting system that integrates with the timesheets? Is the information available in real time?
If not, are your PMs creating budgets in Microsoft Excel? What templates are they using?
Are PMs given all the information that they need to manage their projects?
Do they understand all the information they are receiving? (This is tricky but an important aspect to uncover.)
Do your PMs use their own spreadsheets to understand how their projects are doing (just because you have a system, don't assume they are not doing this)?
Do your PMs get alerts to warn them about excess expenditures on their projects, excessive past due AR balances, or staff charging to projects in excess of the budget?
What is your current process for continuing or stopping work on projects when clients are severely overdue on their receivables? (This affects project profit and cash flow.)
Is there high turnover of staff in the firm or on projects?
Have contingencies been adequately planned for in the budgets?
Are project meetings being tracked, and are they adequately estimated in budgets?
Is your firm hiring the right people?
Do your managers get trained on how to manage staff, and evaluate performance?
Do project management reports indicate any trends or issues with particular managers or clients?

The best way to find out this information about your PMs and their challenges in managing their projects is to ask them, and evaluate data from your system. By interviewing all of your PMs, the good and the bad, you will start to recognize some patterns. While some of the issues may be due to poor behavior of individuals, you will be able to isolate these problems, and gather valuable data about the bottlenecks that exist in the firm, causing projects to run over budget.

Cost of a Long Invoice Cycle and Poor Cash Flow

As I described in Chapter 2, poor cash flow is due to many operational factors, including pursuing the wrong projects, inadequate time management, a long billing cycle, and deficient collection processes. While the cost of money is low these days, and the related cost savings to the bottom line may not be large, poor cash flow puts a strain on your firm, your managers, and your clients.

Analysis of the following practices in your firm will help you determine how cash flow can be improved:

Do you have detailed documented timesheet procedures that every employee is required to follow?
Are your employees filling out their timesheets on a daily basis?
How are timesheet policies enforced to ensure compliance?
Do your PMs rigorously approve timesheets?
Are there a lot of corrections to timesheets during the approval process?
Are there a lot of transfers of time during the billing process?
Are your subcontractors getting you their invoices on a timely basis for your billing cycle?
Are your subcontractor invoices accurate?
How many days does it take to get your bills out the door?
Are you emailing managers their billing reports? What reports are they getting?
Are your billing managers given a deadline to review and return their billing edits?
What is holding up the process of billing managers getting them back to accounting in the required amount of time?

Are you emailing your invoices?
How often are clients disputing invoices?
How often do clients pay the wrong amount?
Does your accounting department review the first invoice with a new client?
Are you calling clients after you send the bills to confirm receipt and ask about any questions?
After how many days overdue are you making collection calls?
How often are collections calls made?
Who is making collections calls?
Are notes recorded about collections calls?
What other collection processes have you put in place?
Are you charging interest on late payment of invoices?
Are you asking for retainers at the beginning of projects?

By evaluating the monthly process and flow of data from the timesheet entry process through the billing review and adjustments, invoicing and follow-up, and collection practices, you will start to see how the invoice and cash collection cycle is being drawn out.

Cost of Inefficient and Non-integrated Systems with Multiple Silos of Redundant Data

While there is no metric we can calculate that helps us analyze the lost dollars from inefficient systems, there are ways of measuring the efficiency of your business practices for your employees. The amount of time spent doing various tasks, measured in man-hours, can be evaluated to determine the savings from integrated systems. The following are the questions that should be asked to determine how well your current systems and processes are performing in your daily operations:

Are there areas where data (such as employee's time or client and project information) is being entered more than once? Places to look at include spreadsheets, outside payroll and HR systems, separate databases, and Microsoft Word or Excel for billing.
Do you have any manual (paper) processes? Document all that exist.
Are you maintaining several databases of client information? Are you sure which is the most accurate? (Be sure to look at how Microsoft Outlook and other systems are being managed.)
Are your separate systems fully integrated? Are there issues with keeping them synced?
Are spreadsheets being prepared? If so, document all the purposes and get copies. Who has access to them and who is responsible for keeping them updated?
Interview your staff to understand typical bottlenecks to getting things completed including paper processes, approvals, and looking for employee and project information.
Is your administrative staff spending time filing paper reports and documents?
How long does it take to find information about clients or projects? Is it automated? Does everyone do it the same way or use the same data?
Do you use paper project file folders?
Do you have a log book for new projects or work orders?
Do you have multiple offices, or virtual or remote employees? If so, are they sharing all of the same information? How is the performance of the systems they are using (slow, fast, etc.)?
Does your IT person have to manage many different servers, applications, licenses and version updates? Is this efficient? Are they current on updates and support?
Is there an IT disaster recovery plan? Are backups being tested regularly?
Survey your employees in each area of the business to determine where they are frustrated and feel that time is wasted.

Based on the answers you get to these questions, you may need to dig deeper and interview more staff. I recommend that you talk to as many people, in as many roles as possible. The more you can help your employees to get their jobs done faster and more accurately, the better positioned your firm will be to grow without having to add more layers of overhead labor.

Cost of Losing a Client

Keeping clients happy and managing client relationships is critical to your firm's success. If you have a small number of clients, it is much easier than if you have a lot of clients. As your firm grows, and you add employees, offices, and projects, the effort to maintain your client relationships often becomes more difficult.

We can all relate to the emotional trauma of learning that a valued client has left your firm for a competitor. While it happens to every business at some point in their history, it is never easy, and often causes deep introspection and evaluation to try and understand what happened. Because new clients are so hard to get these days, it is even more painful than ever before to contemplate losing a client.

In taking an honest look at the experience that your clients have in working with your company, you can start to put some practices and systems in place to improve your firm's responsiveness, work product, and level of responsibility. The following questions and evaluation can help you to determine if you are doing everything possible to keep your clients happy, and avoid the pain of losing a valuable client:

What is the process for inbound client inquiries to your firm?
Is the process documented and followed?
Are new employees coached about the firm's mission, vision and values?
Do you monitor client complaints?
Do you survey clients during and after a project to determine their level of satisfaction with your work? If so, how often?

Do your employees document all of your phone conversations, emails and meetings with clients with notes and action items?
If an employee leaves your firm, how do you transition their knowledge of your clients and projects to the employees taking over the project?
How often do your Principals meet with clients in person?
Are you able to analyze which clients you make the most money from, and which ones you are losing money on?
Do you have clients that consistently do not pay you?
Are there some clients that you have consistent issues with (personality, changes in scope, not paying)? Have you analyzed your profitability on these clients?
Do you treat all clients the same, or do you have special processes for your best clients?
How often do you talk with your clients about business issues outside of your immediate project work?
Do you have clients that give you referrals? How do recognize them for this?
Do you track personal information about your clients such as family members, birthdays, or hobbies and interests?
What percentage of clients give you repeat business?
What do you do above your project responsibilities to help clients?
What other activities do you engage in that add value to your clients' businesses?
Do you have a written plan for retaining existing clients?

In Step 4, we will look at how to develop a plan to use the information you have gathered to find the lost dollars in your business.

STEP 4: DEVELOP A PLAN TO MAKE IMPROVEMENTS

If you have taken the time and made the effort to go through the exercises in Step 3, then you are halfway through the process of finding the lost dollars in your business. While it is not necessary to have addressed all nine areas of your business, it can be extremely eye opening and given you a new perspective on the true operational deficiencies in your business.

With the understanding of the impact of only a 1% change in your key metrics, you should be feeling pretty confident that you have both the analysis, and the motivation to make the changes that are necessary to make. Now it is time to take what you have learned, and formulate a plan for improving your business operations.

You will not be able to just take this plan and implement it in days. Part of the purpose of Step 1 and 2 was to show you where you can get the greatest return on your investment. I recommend that you attack this project in three phases, starting with the metric that will give you the biggest annual return. While you may feel that you need to jump on all of these issues immediately, you will not be as successful doing this, as you will if you take on one problem at a time. Depending on the number of issues that you uncovered, and the degree to which change must be implemented in your company, this entire process can take a few months to a couple years.

In order to develop a plan, the first step is to group all of the information that you uncovered into categories. I recommend you use the categories listed below, however, you may find that this is too detailed or that you require additional categories in order to be able to address the issues appropriately. Take each of the answers to the items that you researched in Step 3, and list the exact problem that you want to fix under one (or more) of the following categories:

- Marketing and Lead Management Processes
- Opportunity and Pipeline Management
- Proposal Preparation
- Time Management and Approvals
- Billing and Collections
- Resource Management and Scheduling

- Project Estimating and Budgeting
- Project Management
- Scope Management and Extra Services
- Multiple databases and redundant data and processes
- Client Relationship Management

After you have documented all of the areas that you want to address under each of these categories, the next step is to develop a strategy to address it. For the most part, the solution to most of these problems is going to be one or more of the following three strategies:

- Implement automated systems
- Improved processes and procedures
- Training and staff development

Next to each of the items, indicate which of the three strategies above you believe would be the most effective. In some cases, such as project management ineffectiveness, you may require all three.

By looking at your issues in their categories, with the solutions listed next to them, you should start to see some patterns and consistency that will enable you to pinpoint the most prolific strategies. For example, if you have excellent systems, but no one is using them very well, you may determine that your documented processes need an overhaul, and your staff needs training. If your systems are poor or non-existent in some areas, or systems are not integrated, you may determine that fixing your systems is the best approach.

The document you have created in this exercise will prove to be a very valuable tool in helping you determine where to focus your attention, what approach to take, and where to start. By focusing on the solutions that are going to help you improve your three metrics, as well as improve cash flow, efficiency in the operations of the business, and ensuring excellent client relationships, you are well on your way to stellar firm performance. In any case, you should see a measurable improvement in your bottom line.

As you look at each of these categories, try to relate each of them back to a specific chapter in the book. For example, if time management processes are a big challenge for your firm (as they are for many of your peers), go back and read Chapter 6 for best practices that you can start to implement. You will probably need to look at your system, processes, and how your employees are actually behaving to determine the best solution to improve each area.

All of the changes that you may decide to make in your business will render a positive end result. All of these categories are intertwined in many ways, and changing one will very often have a positive change on another with little or no effort. As you start to see improvement in your business, your employees should start to feel encouraged, and their productivity should increase. All of these things add up to fewer lost dollars and higher profits.

STEP 5: IMPLEMENT CHANGES

The fifth step is to implement the plan that you have created in Step 4. While this is probably the most time-consuming aspect of this exercise, it is also where you will start to feel that you are really making a difference in your business, instead of just looking at it and getting frustrated.

I tried to provide you with some guidelines of how to go through the process of implementing changes. In Chapter 5, we looked at the concept of business process change, and I recommended a process for how to evaluate your current processes, map out new ones, document them and implement them. Being aware of how your firm deals with change, and putting steps in place to communicate how things are going and why the firm is changing are important.

In Chapter 9 I gave you practical guidelines on how to select and implement software solutions. This is where a great deal of my experience has been over the last 23 years, and I have done my best to give you an unbiased guide to ensure your success. There is not just one answer for which systems are going to be the best for every company.

I do urge you, though, to consider your short- and longer-term goals as you move forward with buying and implementing new systems, and to look at your system as a strategic investment that you are leveraging to accomplish those goals. Software is incrementally less expensive than people. Investing in good software today will save hundreds of thousands in the future in improved efficiency, and improved project profitability.

If you determined that training is the key to finding your lost dollars, then invest in programs that will help your people succeed and elevate your business to another level. As a professional services business, you are only as good as the people that you have. If your people are aligned with your mission, vision, and values, and have the tools and training they need to be successful, you will see improved results and experience less problems. Your clients will also have a better experience working with your firm, which will also increase profitability across the board.

You can evaluate many different types of training, depending on what your employees require, including training in accounting, business management, HR (hiring, recruiting, etc.), project management, vendor-specific software training, and management or executive leadership training. All of these different types of training are available as classroom training, some with Continuing Education Units CEUs available for them, as well as web based training.

I recommend that you check with the trade associations with which you work to get their recommendations. In addition, both ZweigWhite and PSMJ have some excellent training programs that can help take your team to the next level.

I urge you to go back and read the relevant chapters in this book again. Numerous ideas and best practices are outlined to help you develop the right approach to your systems, process, and employee development improvements. In some cases, you may need the help of experts. Investments in bringing in outside help can be well worth the cost savings and allow you to make these changes faster and more effectively.

Step 6: Measure and Track Results

If you have made it this far, then you should be starting to see measureable improvements in your business. At a minimum, I guarantee that you have a great deal of additional insight into your business operations and where money is being lost in them.

The last step in this process is to continue to measure your metrics on a monthly basis and track improvements as you implement the changes. You may see some fluctuation as your changes are taking hold and new processes and systems are being learned. In most cases, you should see measurable improvement in one year's time or less. You should compare your metrics to the goals that you set in Step 2, and determine whether any additional changes need to be put in place in order to accomplish these goals. This is really the easiest step in the process—sit back and watch all of your hard work pay off!

One other recommendation I have is to go through the exercise in Step 3 once a year to track progress with both hard metrics and soft, intangible benefits. The information you uncovered in the process of going through this exercise is invaluable, and true progress in improving all the deficiencies that you may have uncovered could take several years. Revisiting these areas, and talking with your staff about improvements they have seen, or areas where they continue to have frustrations will serve to keep your firm in a state of constant improvement. I believe that constant improvement is necessary these days to ensure a competitive edge and flexibility to avoid the traps detailed in Chapter 1.

If you have followed the 6 steps, or still feel challenged to try and accomplish them on your own, I am here to help. My greatest joy comes from helping business owners make the changes necessary to improve their company performance. My team of industry experts and I can help you accomplish the analysis, planning, and implementation of the solutions uncovered through the 6-step process. To contact me to discuss this further, or to give me feedback on your experience in finding the lost dollars in your business, please e-mail me at JJewell@AECBusiness.com

I will wrap up this book with some quotes from some long-time clients and business owners I know who are sharing their wisdom about what has made them successful. I hope you enjoy their outlook. I would love to hear from you too!

Summary

- In order to find lost dollars in your A&E firm, you will need to take a deep look at nine areas of your business where revenue is not captured, time is wasted, or project profits are lost.

- I have outlined 6 steps that can help you to identify where the most money is lost and develop a plan for sealing up the holes.

- The 6 Steps include:

  1. Calculate your metrics.
  2. Identify where the most money can be found.
  3. Analyze the components that affect that metric.
  4. Develop a plan to make improvements.
  5. Implement changes.
  6. Measure and track your results.

- I have provided a detailed plan to analyze the nine areas of your business and determine whether you need new processes, training, improved systems, or some combination of the three in order to find the money that is being lost in your business.

- Please go to http://AECBusiness.com to learn about some automated tools to help you implement the six steps faster and easier.

Appendix A: Success Quotes from A&E Firm Owners and Principals

Here are some quotes from A&E firm leaders that can provide some insight into what it takes to be successful in this industry:

"Being a leader and making difficult decisions and taking risk is never easy, that is why it is important to keep learning as much as you can about your industry, and then go with your gut feeling.

Our firm has been successful not just by understanding our client needs and expectations and then doing all we can to exceed them, but by building relationships that last longer than the projects."

– Patrick J. Guise, Vice President of Finance, McCormick Taylor, Inc., Philadelphia, Pa. (www.McCormickTaylor.com)

"VIKA owes the lion's share of our success to (1) our commitment to continually strive to delight our clients with ever increasing levels of service, and focus on quality services, provided through our 'practicing principals,' and (2) by demonstrating this philosophy and work ethic to our valued employees."

– John F. Amatetti, P.E., Principal and Charles (Chuck) A. Irish, Jr., PE, LS, LEED AP, Chief Operating Officer, VIKA, McLean, Va. (www.VIKA.com)

"I have developed the instinct to survive, and the determination to not just survive but to thrive. To not be one of the victims is also very strong motivation. So perhaps it is the willingness to expand into markets you have never been in before with a determination to succeed by becoming proactive on behalf of your clients. For me the last 18 months has been all about aligning my business with like-minded people who approach the 'new normal' with a 'How can I help you succeed' attitude. It is no longer about me it is about my clients and networking friends. I will do anything I can to help them with no expectation of anything in return. It is the old 'What goes around comes around' approach. Funny—it seems to be working."

– James (Jim) P. Greenfield, AIA, Greenfield Architects, P.C., Denver, Colo., (www.GreenfieldArchitects.com)

"Success in business requires a tenacity to not only persevere through challenges, but to be resourceful in finding solutions; to be perceived as a 'finisher,' regardless of the obstacles."

– Joseph (Joe) A. Cappuccio, P.E., Senior Vice-President, Rolf Jensen & Associates, Inc., Fairfax, Va., (www.RJAGroup.com)

"Always be there for your client, answer your cell phone whenever it rings and never bullshit them. One person can't do this alone, empowering your people and getting them to adopt this mindset will comfort your clients in times of chaos and will be rewarded with their appreciation when the project is successful.

It's never enough to just do good work. You need to give a client reasons to have a rooting interest in your success. The easiest way to do this is by becoming indispensable to the process by making everyone else around you better. Smart, successful clients realize that they didn't succeed by themselves. If they don't, you shouldn't be working with them."

– John C. Saber, PE, President & CEO, Encon Group, Inc., Kensington, Md., (www.EnconGroup.com)

"Some suggested reasons why we have had success:

- *We strive to be a trusted client partner, rather than simply being a vendor executing a contract or providing an off-the-shelf service.*
- *We focus on innovative yet practical solutions to our client's problems.*
- *We have a diverse set of dedicated employees that maintain an emphasis on the company's success."*

– Richard Rehmann, GISP, President, ARH Associates, Inc., Hammonton, N.J. (www.ARH-US.com)

"As a small business owner of 160 staff providing engineering and environmental consulting for the Federal government, we are diligent regarding financial policy, tracking, and documentation to comply with the applicable Federal Acquisition Regulations (FAR) and client requirements, as well as pass Defense Contract Audit Agency (DCAA) audits. This includes a robust accounting system and associated written policies approved by DCAA to ensure compliance. Strong financial management and approved procedures are key to passing critical DCAA audits in this business. Strong financial and technical management of direct billable costs, indirect expenses, and profit in conjunction with providing quality products on time and within budget has proven critical to a 24-year success record from both a growth and DCAA compliance perspective."

– Donna Baird Lawrence, CEO/President, AGEISS Inc., Evergreen, Colo. (www.AGEISS.com)

"Understanding the market place and the competition, and making changes as required and needed, is necessary for aprofitable company. This is how we are making it in this environment."

– Barthy Setty, Vice-President, Setty & Associates, Ltd., Fairfax, Va. (www.Setty.com)

"As designers for the lively and performing arts, those of us at Martinez+Johnson Architecture have found our greatest satisfaction through our efforts in helping to revitalize and strengthen the social and cultural fabric of cities and towns across the nation. Even more than the theater and concert halls we have created, the friendships and relations we have formed with the residents of the communities in which we have worked, have enriched our lives immeasurably."

– Gary F Martinez, AIA, President, MARTINEZ+JOHNSON Architecture PC, Washington, D.C., (www.MJArchitecture.com)

"Not to be a smart aleck, but being successful in the past wasn't that hard. I'm more focused on how to be successful today and in the future. What worked yesterday isn't working the same today or as we move forward. Changes that managers must address are presenting themselves more rapidly than at any time in the past and that challenge grows as we move ahead."

– Joseph (Joe) Paciulli, L.S., President and CE0, Paciulli, Simmons & Associates, Ltd., Fairfax, Va. (www.PSAltd.com)

"We feel that we have been in the right market sector,in the right location, at the right time. We also have recognized that we are servicing people with diverse needs, risk aversion, and personalities, and that our success depends on mastering customer service, communication and relationship building."

– Jon P. Buzan, PE, President, REPSG - React Environmental Professional Services Group, Inc., Philadelphia, Pa. (www.RESPG.com)

"I am a successful [A&E] business owner due to our determination to foster long-term client relationships. We coach our staff to exceed client expectations. DBI's ability to provide exemplary design services has fueled our business with work from many repeat clients throughout many economic downturns."

– Roseanne Beattie, AIA, DBI Architects, Washington, D.C. (www.dbia.com)

"Our success is due to our exceptional people, a nurturing culture, client focus and 'can do' attitude. In addition we have had a strategy of diversification in multiple markets nationally and internationally. We have transformed ourselves from purely an engineering company to a global 'Design Build Manage company'."

– Zack Shariff, P.E., LEED® AP, CEO, Allen & Shariff Corporation, Columbia, Md. (www.AllenShariff.com)

"Our success stems from a number of key principles in no particular order of importance because all are critical to succeed—be sure you have the right people on the bus, and in the right seat. Then treat your team well—golden handcuffs— if that is the way you are wired. Plan early, plan often, have corporate retreats with all key team leaders. Insure you know what makes your staff 'tick,' do you know how they are wired? Hire a great CEO Coach for a year, hire a great PR firm, be sure your accounting software suits your needs, and surround yourself with very smart consultants in law, lobbying, accounting, etc. Finally, get face time with your clients—do not rely on electronic communication."

– Wes Guckert, President & CEO, The Traffic Group, Inc. Baltimore, Md. (www.TrafficGroup.com)

"First of all, I don't see myself as very successful. I have lots of friends who have started many firms that are wildly successful compared to our firm—they grow faster, have expanded geographically, and make lots more money. What I have done is start a firm (my second) that has been in business for almost 22 years now, which has grown a cadre of dedicated staff whom are collectively known to be the best in our small niche of technical and geographical space. We have done this by a sharp focus on sticking to what we do best, and adding services that improve our ability to help clients pass through the maze of natural and cultural resource regulations created by the Clean Water Act and related federal regulations (National Historic Preservation Act and Endangered and Threatened Species Act), state regulations (through the Chesapeake Bay Act and Stormwater Management Programs) and their related local requirements. I simply love the nature of what we do, and enjoy our firm's culture of simply always striving to be the best; and have the fortune to have many people to work with that share this desire."

– Michael (Mike) S. Rolband, P.E., P.W.S., P.W.D., LEED® AP, President & CEO, Wetland Studies & Solutions, Inc., Gainesville, Va. (www.WetlandStudies.com)

"I can attribute my success as a CEO to my knowledge of marketing, using the strengths of people and compensating for their weaknesses, my ability to articulate our goals and what is our primary objective, but most of all I demand every opportunity and task be approached with a sense of urgency. In this way we can most easily differentiate ourselves from the competition and earn the respect of our customers. The engineers who adopt this philosophy move ahead in our firm. Those who don't may last a long time but more than likely it is the market that really removes them. I just make sure nature's course is not obstructed by weak leadership when the decisions required for growth are obvious."

– Charles F. Hammontree, President/CEO, Hammontree and Associates, Limited, Canton, Ohio (www.Hammontree-Engineers.com)

"Diversify your client base but stick to what you know. I worked for 5 different MPE engineering firms prior to starting WFT Engineering. I quickly noticed that many of these firms stuck to and serviced a small number of clients and would succeed or fail based on the success of one or two clients and or markets. When starting WFT Engineering, I was determined that I was not going to fall into the same trap so I specifically pursued work in the federal and state markets, commercial markets and mission critical markets. The engineering fundamentals don't change based on the client ... only the tools for application."

– Reardon (Sully) D. Sullivan, P.E., President & CEO, WFT Engineering, Rockville, Md., (www.WFTeng.com)

"Long-term success in the A&E business comes from three primary components, each building on the other. You must first deliver design work of the highest quality. This brings clients to your door. Add to that a sincere devotion to delivering excellent client service. This keeps them coming back. The last piece is a dogged commitment to business fundamentals. It's not as much fun as designing creative spaces, of course, but too many talented firms fall down because of sloppy practice habits."

– James (Jim) Myers, AIA, Managing Principal, GTM Architects, Bethesda, Md., (www.GTMarchitects.com)

"Civil engineering and land planning are natural to me—I love to see things being built. Surround yourself with thinking and creative people and put them in roles for success. Also when monetary success comes to your company you have to share."

– Mark D. Smith, PE, LS, President, Greenway Engineering, Winchester, Va. (www.GreenwayEng.com)

Appendix B: Contributors (in order of appearance)

Thanks to all of the firm leaders and industry consultants that contributed their wisdom and expertise to this book:

Mark C. Zweig, Founder and CEO, ZweigWhite, Fayetteville, Ark., (www.ZweigWhite.com)

Frank Stasiowski, FAIA, Founder and CEO, PSMJ | Resources, Inc., Newton, Mass, (www.PSMJ.com)

Ed Friedrichs, Founder, Friedrichs Group, San Francisco, Calif., (www.FriedrichsGroup.com)

Christine Brack, Principal, ZweigWhite, Chicago, Ill., (www.ZweigWhite.com)

Bob Gillcrist, Sr. Client Executive, Deltek, Herndon, Va., (www.Deltek.com)

Derrick Smith, Senior Vice-President, Mackay Sposito, Vancouver, Wash., (www.MacKaySposito.com)

John C. Saber, PE, President & CEO, Encon Group, Inc., Kensington, Md., (www.EnconGroup.com)

Robert Skepton, Chief Financial Officer, Hillis-Carnes Engineering Associates, Annapolis Junction, Md., (www.HCEA.com)

Ted Maziejka, LEED AP, Consultant, ZweigWhite, (www.ZweigWhite.com)

Mike Phillips, AIA, President, Phillips Architecture, Atlanta, Ga., (www.PhillipsArch.com)

Ron Worth, President and CEO, Society for Marketing Professional Services (SMPS), Alexandria, Va., (www.SMPS.org)

Lee W. Frederiksen, Ph.D., Managing Partner, Hinge Marketing, Herndon, Va., (www.HingeMarketing.com)

Mark Amtower, Founder, Government Market Master, and Author, Columbia, Md., (http://governmentmarketmaster.com/)

Brenda S. Stoltz, CEO, Ariad Partners, Leesburg, Va., (www.AriadPartners.com)

Brian Bass, Senior CRM Consultant, Acuity Business Solutions, (www.AcuityBusiness.com)

Tim Klabunde, Marketing Director, Timmons Group, Richmond, Va., (www.Timmons.com)

Nancy Usrey, FSMPS, CPSM, Design-Build Team at HNTB, Principal Consultant at Partners Usrey, Dallas, Tex., (www.HNTB.com)

Bob Stalilonis, Senior Solutions Architect, Deltek, Herndon, Va., (www.Deltek.com)

Deborah Gill, CPA, CDFA, Controller, Clark Nexsen, Norfolk, Va., and 2012 / 2013 President of Society for Design Administration (SDA), (www.ClarkNexsen.com)

Anthony (Tony) J. Vitullo, CPA, CGMA, CFO, Cooper Robertson Partners, New York, N.Y., (www.CooperRobertson.com)

June Pride, Senior Accountant, McCormick Taylor, Philadelphia, Pa., (www.McCormickTaylor.com)

Edward J. Kroman, III, Director of Finance, Summer Consultants, Inc., McLean, Va., (www.SummerConsultants.com)

Robert (Bob) Johansen, Director of Information Analytics, Leo A. Daly, Omaha, Neb., (www.LeoADaly.com)

George E. Christodoulo, PC, LAWSON & WEITZEN, LLP, Boston, Mass., (www.Lawson-Weitzen.com)

Jonathan Voelkel, Director of Special Projects, Rusk O'Brien Gido + Partners, Falls Church, Va., (www.rog-partners.com)

Jay Appleton, Principal, Kitchen & Associates, Collingswood, N.J., (www.KitchenandAssociates.com)

David A. Stone, CEO, Stone & Company, Everett, Wash., (www.StoneandCompany.net)

Larry G. Kirk, CPA, Vice-President of Finance, AES Consulting Engineers, Williamsburg, Va., (www.AESVA.com)

Nick Castellina, Research Analyst, Enterprise Applications, Aberdeen*Group,* Boston, Mass., (www.Aberdeen.com)

Sheldon Needle, Founder of CTS Guides, Rockville, Md., (www.CTSGuides.com)

George J. Nagy, Sr., Telemus Group, LLC, Chevy Chase, Md., (www.TerminusGroupllc.com)

Guy Timberlake, Founder and Chief Visionary of the American Small Business Coalition (ASBC), (www.theASBC.org)

In addition to the experts that contributed to the content for the book, I would like to recognize the successful A&E firm owners, many of them my long-term and most valued clients, who offered their thoughts on what has made them successful:

Patrick J. Guise, CEO, McCormick Taylor, Inc., Philadelphia, Pa., (www.McCormickTaylor.com)

John F. Amatetti, P.E., Principal and Charles (Chuck) A. Irish, Jr., PE, LS, LEED AP, Chief Operating Officer, VIKA, McLean, Va., (www.VIKA.com)

James (Jim) P. Greenfield, AIA, President & CEO, GREENFIELD ARCHITECTS, p.c., Denver, Colo., (www.GreenfieldArchitects.com)

Joseph (Joe) A. Cappuccio, P.E., Senior Vice-President, Rolf Jensen & Associates, Inc., Fairfax, Va., (www.RJAgroup.com)

John C. Saber, PE, President & CEO, Encon Group, Inc., Kensington, Md., (www.EnconGroup.com)

Richard Rehmann, GISP, President, ARH Associates, Inc., Hammonton, N.J., (www.ARH-US.com)

Donna Baird Lawrence, CEO/President, AGEISS Inc., Evergreen, Colo., (www.AGEISS.com)

Barthy Setty, Vice-President, Setty & Associates, Ltd., Fairfax, Va., (www.Setty.com)

Gary F Martinez, AIA, President, MARTINEZ+JOHNSON Architecture PC, Washington, D.C.

Joseph (Joe) Paciulli, L.S., President and CE0, Paciulli, Simmons & Associates, Ltd., Fairfax, Va., (www.PSAltd.com)

Jon P. Buzan, PE, President, REPSG - React Environmental Professional Services Group, Inc., Philadelphia, Pa., (www.REPSG.com)

Roseanne Beattie, AIA, DBI Architects, Washington, D.C., (www.dbia.com)

Zack Shariff, P.E., LEED® AP, CEO, Allen & Shariff Corporation, Columbia, Md., (www.AllenShariff.com)

Wes Guckert, President & CEO, The Traffic Group, Inc. Baltimore, Md., (www.TrafficGroup.com)

Michael (Mike) S. Rolband, P.E., P.W.S., P.W.D., LEED®
AP, President & CEO, Wetland Studies & Solutions, Inc.,
Gainesville, Va., (www.WetlandStudies.com)

Charles F. Hammontree, President/CEO, Hammontree and
Associates, Limited, Canton, Ohio, (www.Hammontree-Engineering.com)

Reardon (Sully) D. Sullivan, P.E., President & CEO, WFT
Engineering, Rockville, Md., (www.WFTEng.com)

James (Jim) Myers, AIA, Managing Principal, GTM Architects,
Bethesda, Md., (www.GTMArchitects.com)

Mark D. Smith, PE, LS, President, Greenway Engineering,
Winchester, Va., (www.GreenwayEng.com)

ABOUT THE AUTHOR

June Jewell has over 28 years of business management consulting experience in the areas of financial best practices and business management systems. Since 1989, June has worked with over 700 A&E firms across the U.S. to help them transform their business management processes and systems to increase project profitability, and improve the productivity of their teams.

June is the CEO of a successful consulting practice, and a sought after speaker at industry focused events providing best practices guidance and innovative ideas for business growth. She has spoken for a wide range of industry organizations including ACEC, AIA, SMPS, ZweigWhite, Design and Construction Network (DCN), CREW, and at Deltek Insight.

For more information on how June can help you find the lost dollars in your business, go to www.FindTheLostDollars.com or email her at June@FindTheLostDollars.com.

Are you Ready to Find the Lost Dollars in Your Business?

Feel overwhelmed with all the areas of your business that need your attention? Let me help you take the mystery out of analyzing the numbers.

Calculate the ROI (lost dollars) from improving different parts of your business

Learn where to focus your attention to get the biggest financial impact

Develop a strategy for finding the lost dollars in your business!

Learn about our online Business Management Assessment Tool and training for Project Managers to teach them business skills

please go to

http://AECBusiness.com

to see a demo of our

Scope Creep training course

CPSIA information can be obtained
at www.ICGtesting.com
Printed in the USA
FFOW03n0423201215
19545FF

9 780988 382428